I0055199

Photochemistry: A Conceptual Approach

Photochemistry: A Conceptual Approach

Carol Ramsey

WILLFORD PRESS

www.willfordpress.com

Published by Willford Press,
118-35 Queens Blvd., Suite 400,
Forest Hills, NY 11375, USA

Copyright © 2022 Willford Press

This book contains information obtained from authentic and highly regarded sources. All chapters are published with permission under the Creative Commons Attribution Share Alike License or equivalent. A wide variety of references are listed. Permissions and sources are indicated; for detailed attributions, please refer to the permissions page. Reasonable efforts have been made to publish reliable data and information, but the authors, editors and publisher cannot assume any responsibility for the validity of all materials or the consequences of their use.

Trademark Notice: Registered trademark of products or corporate names are used only for explanation and identification without intent to infringe.

ISBN: 978-1-64728-009-3

Cataloging-in-Publication Data

Photochemistry : a conceptual approach / Carol Ramsey.
    p. cm.
Includes bibliographical references and index.
ISBN 978-1-64728-009-3
1. Photochemistry. 2. Chemistry, Physical and theoretical. I. Ramsey, Carol.
QD708.2 .P46 2022
541.35--dc23

For information on all Willford Press publications
visit our website at www.willfordpress.com

WILLFORD PRESS

Table of Contents

Permissions

Index

Preface

It is with great pleasure that I present this book. It has been carefully written after numerous discussions with my peers and other practitioners of the field. I would like to take this opportunity to thank my family and friends who have been extremely supporting at every step in my life.

The branch of chemistry which is concerned with the chemical effects of lights is known as photochemistry. It generally deals with chemical reactions that are caused by the absorption of ultraviolet, visible light and infrared radiation. The most common photochemical reactions include bioluminescence, photoresist, photosynthesis, vision, toray, photodegradation and photodynamic therapy. Some of the major concepts and laws within this field are Grotthuss–Draper law, Stark-Einstein law, Kasha's rule and Hund's rule of maximum multiplicity. There are various areas of study within this field such as organic photochemistry, inorganic photochemistry and organometallic photochemistry. This book unfolds the innovative aspects of photochemistry which will be crucial for the progress of this field in the future. Those in search of information to further their knowledge will be greatly assisted by this book. Coherent flow of topics, student-friendly language and extensive use of examples make it an invaluable source of knowledge.

The chapters below are organized to facilitate a comprehensive understanding of the subject:

Chapter – What is Photochemistry?

The branch of chemistry which aims at studying the chemical effects of light is referred to as photochemistry. It describes the chemical reaction which is caused by the absorption of ultraviolet or infrared radiation. This is an introductory chapter which will introduce briefly all the significant aspects of photochemistry.

Chapter – Photochemical Reactions

A photochemical reaction takes place when a molecule comes in contact with the light. Luminescence, phosphorescence, fluorescence, photodegradation, photosensitization, photodissociation, etc. are some of the concepts that fall in this domain. The concepts elaborated in this chapter will help in gaining a better perspective about these photochemical reactions.

Chapter – Photosynthesis

The process used by plants and other organisms in which light energy is used to produce glucose from carbon dioxide and water is referred to as photosynthesis. This glucose is then converted into adenosine triphosphate to provide energy. This chapter has been carefully written to provide an easy understanding of photosynthesis.

Chapter – Ultraviolet Radiation

The electromagnetic radiation with a wavelength shorter than that of visible light but longer than the soft X-rays is known as ultraviolet radiation. Some of the other aspects of ultraviolet radiation are UV curing, ultraviolet germicidal irradiation, UV degradation and UV pinning. This chapter closely examines these aspects related to ultraviolet radiation to provide an extensive understanding of the subject.

Chapter – Diverse Aspects of Photochemistry

Some of the vital aspects of photochemistry are photocatalysis, photoelectrochemical processes, photovoltaic effect, photovoltaic catalysis and photoelectric effect, etc. It also includes various types of photovoltaic cells such as thin-film solar cell, organic solar cell and hybrid solar cell. This chapter discusses in detail all these diverse aspects of photochemistry.

Carol Ramsey

1

What is Photochemistry?

The branch of chemistry which aims at studying the chemical effects of light is referred to as photochemistry. It describes the chemical reaction which is caused by the absorption of ultraviolet or infrared radiation. This is an introductory chapter which will introduce briefly all the significant aspects of photochemistry.

Photochemistry, a sub-discipline of chemistry, is the study of the interactions between atoms, molecules, and light (or electromagnetic radiation). The chemical reactions that take place through these interactions are known as photochemical reactions. Examples of photochemical reactions are photosynthesis in plant cells and light-induced changes that take place in the eye. In addition, photochemical reactions are important in photography, dye bleaching, and television displays.

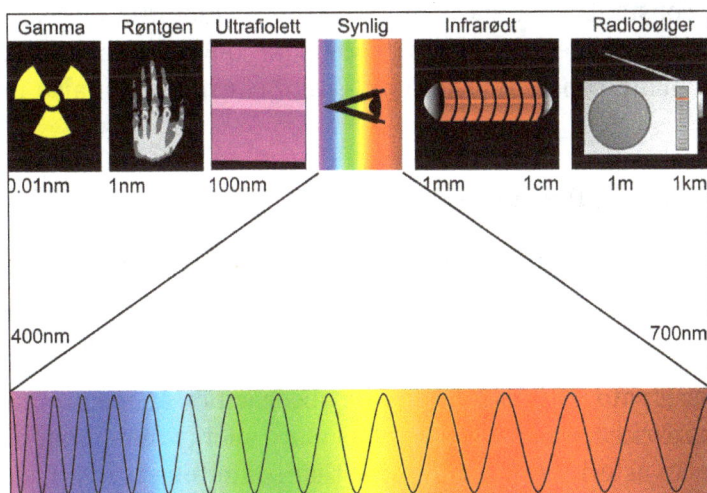

Significant regions of the electromagnetic spectrum, with their approximate wavelengths. The bottom part of the illustration shows an expanded version of the visible spectrum.

Reactions Activated by Light

A photochemical reaction may be thought of as a reaction ignited by the absorption of light. Normally, a reaction (not just a photochemical reaction) occurs when the molecules involved gain the activation energy necessary to undergo change. For example, for the combustion of gasoline (a hydrocarbon) to produce carbon dioxide and water, activation energy is supplied in the form of heat or a spark. In the case of photochemical reactions, light provides the activation energy. The absorption of light by a reactant elevates the reactant to a higher energy state, or excited state, and the process is called "photoexcitation."

The absorption of a photon of light by a reactant molecule may permit a reaction to occur not just by bringing the molecule to the necessary activation energy, but also by changing the symmetry of the molecule's electronic configuration, enabling an otherwise inaccessible reaction path.

A substance that absorbs radiation and transfers energy to the reactant is called a "photosensitizer." When a photoexcited state is deactivated by a chemical reagent, the process is called "quenching."

Regions of the Electromagnetic Spectrum

The electromagnetic spectrum is broad, but photochemists find themselves working with several key regions:

- Visible Light: 400–700 nanometer (nm) wavelength range.

- Ultraviolet: 100–400 nm wavelength range.

- Near Infrared: 700–1000 nm wavelength range.

- Far infrared: 15–1000 micrometer (μm) wavelength range.

Units and Constants

Like most scientific disciplines, photochemistry utilizes the SI, or metric, measurement system. Important units and constants that show up regularly include the meter (and variants such as centimeter, millimeter, micrometer, and nanometer), seconds, hertz, joules, moles, the gas constant R, and the Boltzmann constant. These units and constants are also integral to the field of physical chemistry.

Steps in Photochemical Processes

The excited state: A photochemical change takes place in two steps. Imagine that a light beam is shined on a piece of gold. The light beam can be thought of as a stream of photons, tiny packages of energy. The energy of the photon is expressed by means of the unit $h\nu$.

When a photon strikes an atom of gold, it may be absorbed by an electron in the gold atom. The electron then becomes excited, meaning that it has more energy than it did before being hit by the photon. Chemists use an asterisk (*) to indicated that something is in an excited state. Thus, the collision of a photon with an electron (e) can be represented as follows:

$$e + h\nu \rightarrow e^*$$

Once an electron is excited, the whole atom in which it resides is also excited. Another way to represent the same change, then, is to show that the gold atom (Au) becomes excited when struck by a photon:

$$Au + h\nu \rightarrow Au^*$$

Emission of energy: Electrons, atoms, and molecules normally do not remain in an excited state for very long. They tend to give off their excess energy very quickly and return to their original state. When they do so, they often undergo a chemical change. Since this change was originally made possible by absorbed light energy, it is known as a photochemical change.

The formation of ozone is just one example of the many kinds of photochemical changes that can occur. When solar energy breaks an oxygen molecule into two parts, one or both of the oxygen atoms formed may be excited. Another way to write the very first equation above is as follows:

$$O_2 + h\nu \rightarrow O^* + O$$

The excited oxygen atom (O*) then has the excess energy needed to react with a second oxygen molecule to form ozone:

$$O^* + O_2 \rightarrow O_3$$

Another way for an excited atom or molecule to lose its energy is to give it off as light. This process is just the reverse of the process by which the atom or molecule first became excited. If the atom or molecule gives off its excess energy almost immediately, the material in which it is contained glows very briefly, a process known as fluorescence. If the excess energy is given off more slowly over a period of time, the process is known as phosphorescence. Both fluorescence and phosphorescence are examples of the general process of light emission by excited materials known as luminescence.

Photochemistry Laws

The first law of photochemistry states that only the light absorbed by a molecule can produce photochemical modification in the molecule. the term "molecule" is broadly defined and includes also atoms, radicals, etc. The law emphasizes the importance of light absorption by the molecule involved in the primary photoprocess, which is a chemical reaction or a physical process involving directly excited species. All aspects and consequences of this law must be considered for quantitative analysis of a photoreaction. This is generally taken for granted, but the frequent practice of comparing photochemical kinetic traces for different molecules without referring to their absorbance suggest that it is ignored more often than one may assume.

The second law of photochemistry was formulated at the beginning of 20th century when the quantum theory was just emerging. It states that one molecule is excited for each quantum of radiation absorbed. In other words, the absorption of light by a molecule is a one-photon process. Therefore for a primary photoprocess only one molecule reacts for each photon absorbed. Typically several competing processes occur in the excited state. In this case, the second law can be reformulated as: the sum of the quantum yields for the primary processes must be unity.

It has taken about 20 years and the development of quantum mechanics to predict two-photon absorption The first experimental observation of the two-photon absorption was made when lasers were developed. Further development in laser technology made almost routine the generation of ultrashort light pulses (10-12 - 10-15 s). Such ultrafast lasers made possible not only experimental study, but also the broad application of multiphoton processes. Multiphoton fluorescence is widely used in imaging of cells and biological tissues. Multiphoton photochemistry recently received attention as a tool for time-resolved studies of important biological processes.

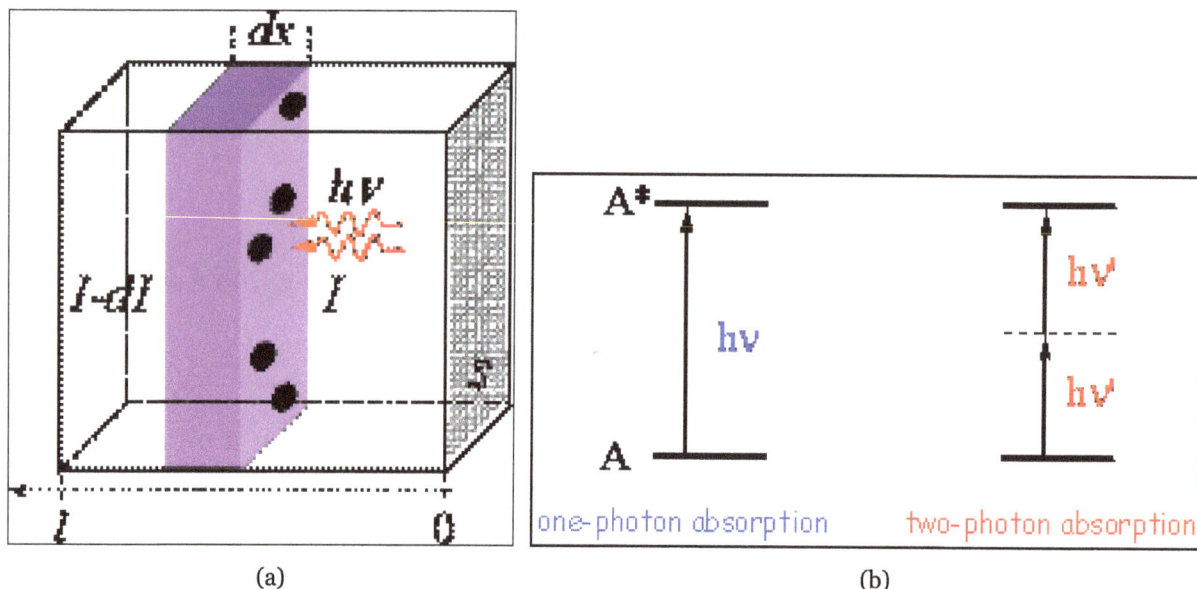

(a)   (b)

(a) Schematic illustration of the light absorption by a rectangular sample. To a first approximation, molecules can be considered as opaque disks whose average cross-sectional area, σ, in cm2 molecule-1 represents the effective area that is impermeable for photons of a certain wavelength. We may consider an infinitesimal slab, dx, of a rectangular sample with a cross-section, S, which is equal to that of the light beam. The average intensity of light entering the slab is denoted I, and expressed in photon s-1. The intensity absorbed in the slab can be written as: $dI = -I\sigma N dx$, , where N is the concentration in molecule cm-3. Integrating this equation from 0 to l (sample length in cm) we obtain the Beer-Lambert law for one-photon absorption: $\ln(I_0/I) = \sigma N l$. If the concentration C is expressed in mol L-1 then the natural logarithm is usually substituted with the decimal one and the cross-section is replaced with the decimal molar absorptivity $\varepsilon \dfrac{\sigma N_A}{1000 \ln 10}$.

Thus, we obtain: $\log(I_0/I) = \varepsilon C l$.

(b) Energy diagrams for one- and two-photon absorption. The average rate of n-photon absorption per molecule in photons s1 molecule-1 can be approximated as: $w_n = \sigma_n I_0^n/S^n$, where σ_n is the cross-section of n-photon absorption, I is the average intensity in photon s-1, and S is the cross-section in cm² of the laser beam entering the sample. For two-photon absorption the cross section, σ_2, has dimension of cm4 s photon-1 molecule -1 and it is often expressed in GM, where 1 GM = 10-50 cm4 s photon-1 molecule-1. The unit was selected to honor Maria Göppert-Mayer who first predicted multiphoton absorption. The measured absorption rate W_n is the number of photons absorbed per s: $W_n = (I_0/I) = w_n N V_{ex}$, where Vex is the excitation volume. For one-photon absorption we obtain: $W_1 = w_1 N V_{ex} = I_0 \sigma_1 N V_{ex}/S = I_0 \sigma_1 N l$. This expression corresponds to the Beer-Lambert law limit for low absorption: $W_1 = (I_0 - I) = I_0\left(1 - e^{-\sigma_1 N l}\right) \approx I_0 \sigma_1 N l$.

Photochemical Kinetics

Quantum yield is the major characteristics of a photochemical reaction. The quantum yield, also called the quantum efficiency, is defined as the number of events occurring per photon absorbed. These events might be related to physical processes responsible for energy dissipation, but they also might be related to molecules of a chemical product formed upon photoirradiation. Generally, the (total) quantum yield of a photoreaction, ϕ, is:

$$\phi = \frac{\text{number of molecules undergoing the reaction of interest}}{\text{number of photons absorbed by the photoreactive substance}}$$

Equation above would define the quantum yield of product formation, ϕ_p, if the number of product molecules would be determined. If the two numbers in equation above are measured per time and volume unit then the quantum yield is expressed in terms of rates:

$$\phi = \frac{\text{rate of the reaction of interest}}{\text{rate of light absorption by the photoreactive substance}}$$

The latter quantity is also referred to as the differential quantum yield. Notice that these two definitions of the quantum yield agree only if the yield is constant during the course of the reaction.

Eqs. $\phi = \dfrac{\text{number of molecules undergoing the reaction of interest}}{\text{number of photons absorbed by the photoreactive substance}}$ and

$$\phi = \frac{\text{rate of the reaction of interest}}{\text{rate of light absorption by the photoreactive substance}}$$

indicate that two separate measurements may be required to determine a quantum yield. In the simplest set-up, a reaction cell is mounted in a fixed position relative to the light source. The cell is charged with the sample of interest and irradiated. Photochemical conversion is determined with a suitable experimental technique (spectroscopy, chemical analysis, etc.). Afterwards, the cell is replaced with an actinometer, which is also irradiated. Before describing how actinometers work, it is important to say again that the amount of the radiation absorbed by the sample, rather than the total amount of light, has to be quantified.

An actinometer is a physical device or chemical system, which is used to determine the number of photons in a light beam. Physical devices convert the energy of absorbed photons into another energy form, which may be easily quantified. The devices that operate by converting photon energy into heat represent 'primary' standards of actinometry. Other physical devices and chemical systems must be calibrated. Chemical actinometers are photoreactive mixtures with well-established photochemistry and known quantum yields. Two representative systems for liquid-phase actinometry are potassium ferrioxalate system and azobenzene system. In both cases, the photoconversion is monitored spectrophotometrically. It is interesting that the most frequently used ferrioxalate system has relatively complex chemistry. Its description in the textbooks hardly goes beyond the statement that Fe(III) is reduced and oxalate is simultaneously oxidized upon photoirradiation. In contrast, the photochemistry of azobenzene is extremely simple. The isomerization reaction proceeds cleanly in both directions and the solution may be regenerated and reused many times.

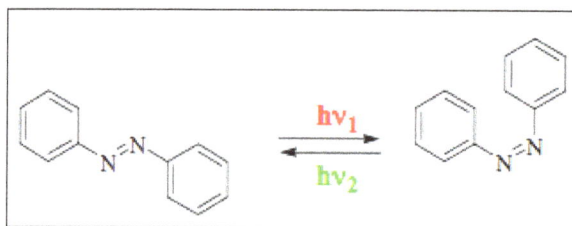

A chemical reaction is just one of multiple routes to the loss of excitation. The light absorption produces an excited-state species that inevitably loses its energy through various deactivation mechanisms. To highlight essential features of photochemical kinetics, we will analyze the simplest system with multiple pathways of deactivation that are characterized by the rate constants corresponding to unimolecular irreversible processes. In the present context, a clear-cut distinction between photophysical processes and a single photochemical reaction will be made. It is assumed that one-photon absorption leads to the direct population of the reactive singlet excited state.

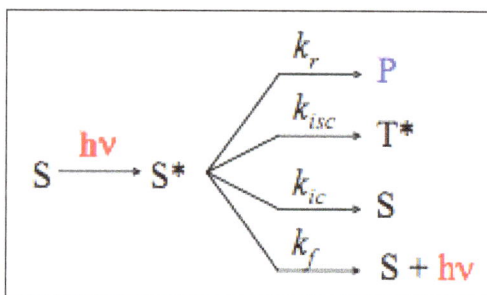

Kinetic scheme for a simple system with a photoreactive singlet state.

The rate constants k_f, k_{ic}, and k_{isc} refer to spontaneous emission, internal conversion and intersystem crossing, respectively. The rate constant k_r corresponds to a chemical reaction. Assuming that population of S* by light absorption is characterized by the constant rate, W, in mol s-1, and the steady-state approximation can be applied to the excited species we obtain the expression for the quantum yield of the photochemical reaction ϕ_0 :

$$0 = \frac{d[S^*]}{dt} = W - \left(k_f + k_{ic} + k_{isc} + k_r\right)[S^*]$$

$$\phi_0 = \frac{k_r[S^*]}{W} = \frac{k_r}{k_f + k_{ic} + k_{isc} + k}$$

Equation $0 = \frac{d[S^*]}{dt} = W - \left(k_f + k_{ic} + k_{isc} + k_r\right)[S^*]$ was solved to obtain an expression for W, which

was then used in equation $\phi_0 = \frac{k_r[S^*]}{W} = \frac{k_r}{k_f + k_{ic} + k_{isc} + k}$.]

Notice that the sum of all the quantum yields is equal to unity, as stated by the second law of photochemistry. The quantum yield for the photoreaction can be interpreted as the fraction of singlet excited molecules that undergo chemical transformation, i.e., the ratio of the number of molecules that react to the total number of S*. Because there always exists several routes to the loss of excitation, the quantum yield rather than the absolute rate constant must be used to compare the efficiencies of photochemical conversion for different reactive systems.

The steady-state approximation is inapplicable under conditions of time-dependent excitation. If a very short laser pulse is used to produce the excited species (so-called δ-pulse excitation) the light absorption rate W can be neglected and equation $0 = \dfrac{d[S*]}{dt} = W - \left(k_f + k_{ic} + k_{isc} + k_r\right)[S*]$ is easily integrated:

$$[S*] = [S*]_0 \exp\left(-\left(k_f + k_{ic} + k_{isc} + k_r\right)t\right) = [S*]_0 \exp\left(-t/\tau_0\right)$$

where τ_0 is the observed lifetime of the singlet excited state. $\tau_0 = \left(k_f + k_{ic} + k_{isc} + k_r\right)^{-1}$

The observed lifetime is an average quantity defined for a large ensemble of the excited molecules. It can be measured with any experimental technique that is capable of detecting the excited species. To take an example, time-resolved fluorescence gives a convenient way of measuring τ_0 provided that certain experimental conditions are fulfilled. Fluorescence detection relies on the photocurrent signal, which is linearly proportional, within certain limits, to the total number of photons emitted. The number of photons, in its turn, is proportional to the number of molecules in the singlet excited state, because individual molecules have time-independent probability to emit light. The fluorescence intensity measured as a function of time depends therefore on the concentration of S*, which is given by equation $[S*] = [S*]_0 \exp\left(-\left(k_f + k_{ic} + k_{isc} + k_r\right)t\right) = [S*]_0 \exp\left(-t/\tau_0\right)$. In contrast to the observed lifetime, the radiative lifetime $\tau_f = 1/kf$ corresponds to the fluorescence decay rate in the absence of any other deactivation processes.

Light should be considered as one of the reactants in photochemical reactions. Therefore, 'effective concentration' of photons, which is given as the number of light quanta absorbed by the photoreactant, needs to be specified when one compares concentration-time profiles for photochemical conversion. In contrast, the efficiency of a thermal reaction can be visualized by plotting the normalized concentration of the reactant or product against time. To clarify this point we need to analyze the photochemical kinetics in more detail. According to the reaction scheme shown in Scheme, the rate of the product formation is:

$$\frac{d[P]}{dt} = k_r[S*] = \phi_0 W$$

Inasmuch as concentrations are determined spechtrophotometrically, it is useful to rewrite $\dfrac{d[P]}{dt} = k_r[S*] = \phi_0 W$ in terms of absorbances:

$$\frac{dA'}{dt} = -\phi_0 \varepsilon_S I_0 \left(A' - A'_\infty\right)\frac{1-10^{-A}}{A}$$

where $A = \left(\varepsilon_S[S] + \varepsilon_p[P]\right)l$ and $A' = \left(\varepsilon'_S[S] + \varepsilon'_p[P]\right)l$ refer to absorbance at the irradiation and observation wavelength, respectively. The absorbance, A'_∞, is measured at the observation wavelength and infinite time, i.e., after complete conversion to the product. In the case of very weak absorption ($A \ll 1$), $\dfrac{dA'}{dt} = -\phi_0 \varepsilon_S I_0 \left(A' - A'_\infty \right) \dfrac{1 - 10^{-A}}{A}$ is easily integrated and experimental data are linearized in the coordinates corresponding to the following equation:

$$\log \frac{A'_0 - A'_\infty}{A' - A'_\infty} = \phi_0 \varepsilon_S I_0 t$$

If we assume that the wavelength where only the reactant absorbs was selected for observation then absorbance in $\log \dfrac{A'_0 - A'_\infty}{A' - A'_\infty} = \phi_0 \varepsilon_S I_0 t$ can be replaced with the reactant concentration. Now a simpler equation, which looks very similar to the rate equation for the first-order thermal reaction can be obtained:

$$\log \frac{[S]_0}{[S]} = \phi_0 \varepsilon_S I_0 t$$

In contrast to thermal reactions, the proportionality coefficient $\phi_0 \varepsilon_S I_0$ in $\log \dfrac{[S]_0}{[S]} = \phi_0 \varepsilon_S I_0 t$ is not just a rate constant independent of the initial concentration, but a complex quantity depending on three parameters. Therefore, the time dependence of the reactant concentration cannot be directly used to compare photoreactivity of different molecules, or even the same molecule if it was measured under different irradiation conditions. We could say that we need to know not only the reactant concentration but also effective 'light concentration', $\varepsilon_S I_0$, in order to analyze photochemical systems. Even if we use the same light source for two systems we cannot directly compare results unless we know how much light was absorbed by each system. Figure shows simulated concentration profiles for two systems that realize the same photochemical reaction S --> P, but differ in spectral parameters and quantum yields. This plot shows how misleading could be a comparison of relative concentrations plotted against time for photochemical reactions if the system is not completely specified.

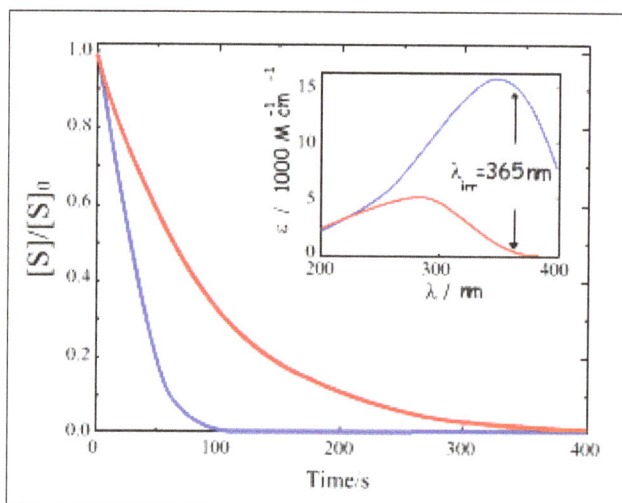

Time profiles for the normalized concentrations of two compounds undergoing an irreversible first-order photoreaction with a quantum yield of 0.1 and 1.0. Solutions containing these compounds at the same initial concentrations were irradiated with the same mercury lamp equipped with a 365 nm narrow-band filter. Which line, blue or red, corresponds to the molecule with the higher quantum yield (more photoreactive)? This question can only be answered when the absorbances at the irradiation wavelength are compared. The substance corresponding to the blue curve has 60 times larger absorbance at 365 nm, which is responsible for faster conversion despite the 10 times lower quantum yield for its photoreaction. As to the question, the correct answer is that the red line corresponds to the molecule with the photoreaction quantum yield of 1.0. However, an extremely weak absorption at 365 nm results in a relatively slow phototransformation of this compound.

In a typical photochemical experiment the concentration of the excited species and transients is negligible in comparison to that of the ground-state species (S and P). Assuming that T* is not reactive and initially we have only the reactant at the concentration $[S]_0$ we can write:

$$[S]_0 \approx [P]+[S].$$

Assuming that light absorption by all transient can also be neglected we can rewrite $\dfrac{d[P]}{dt} = k_r[S^*] = \phi_0 W$ as follows:

$$\frac{d[P]}{dt} \approx -\frac{d[S]}{dt} = \phi_0 W\big([S],[P],t\big)$$

where W([S],[P],t) is the rate of light absorption by the reactant which is a function of time t and concentrations both the reactant and product. To obtain the expression for W we will use the Beer-Lambert law, and the fact the absorbances of components in a mixture add up together:

$$W\big([S],[P],t\big) = I_0 \frac{\varepsilon_s[S]}{\varepsilon_S[S]+\varepsilon_P[P]}\Big(1-10^{-(\varepsilon_S[S]+\varepsilon_P[P])l}\Big)$$

Here I_0 is the intensity of monochromatic light entering the sample expressed in mol L-1s-1, ε_s and ε_p are molar absorptivities of the reactant and product (M-1 cm-1), and l is the optical path (cm). The course of a photochemical reaction is often monitored spectrophotometrically at wavelength(s) different from the irradiation wavelength. By using the Beer-Lambert law we may write for the absorbance measured at the irradiation and observation wavelength at time t: $A = \big(\varepsilon_S[S]+\varepsilon_p[P]\big)l$ and $A' = \big(\varepsilon'_S[S]+\varepsilon'_P[P]\big)l$. By using the absorbance, $A'_\infty = \varepsilon'_P[P]l = \varepsilon'_P[S]_0 l$,

measured at the observation wavelength and infinite time and the three equations shown above we obtain $\dfrac{dA'}{dt} = -\phi_0\varepsilon_S I_0\big(A'-A'_\infty\big)\dfrac{1-10^{-A}}{A}$. In the general case, $\dfrac{dA'}{dt} = -\phi_0\varepsilon_S I_0\big(A'-A'_\infty\big)\dfrac{1-10^{-A}}{A}$ cannot be integrated. But it can be easily solved for very low and very high absorbance, and also for a special case when the product does not absorb at the irradiation wavelength, $\varepsilon_P = 0$. In the later case we obtain:

$$\log\frac{10^X-1}{A^{X_0}-1} = -\phi_0\varepsilon_S I_0 t$$

where,

$$X = \frac{\varepsilon_S}{\varepsilon'_S - \varepsilon'_P}\left(A' - A'_\infty\right) \text{ and } X_0 = \frac{\varepsilon_S}{\varepsilon'_S - \varepsilon'_P}\left(A'_0 - A'_\infty\right)$$

Theoretical Models of Photochemical Reactions

Within the Born-Oppenheimer approximation, potential energy surfaces govern nuclear motion and, therefore, chemical reactivity. However, in studying photochemistry it is also good to keep in mind that this is just an approximation, which is not automatically valid for all possible geometries and experimental conditions. A comprehensive picture of nuclear dynamics can be obtained from the time-dependent Schrödinger equation. However, a detailed account of nuclear motion can also be inferred from classical trajectories for a point moving without friction on the potential energy surface. The moving point may represent a chemically reactive system which consists of one or several molecular species. In the latter case one considers all reactants as a "supermolecule". The forces acting on the nuclei are given by minus the gradient of the potential (electronic energy) at this point. Recall that the gradient for a function of many variables is a vector formed by the first derivatives with respect to each of the variables.

Points on the surface that are characterized by the gradient vector of zero length are called stationary points. Their location is of primary importance for chemical reactivity. The nature of stationary points is determined by the secondary derivatives, the so-called Hessian matrix. If all the eigenvalues of this matrix are positive, the point is a minimum, which can be assigned to a reactant, product or intermediate. A first-order saddle point has all positive eigenvalues except for one, which is negative. It means that it is a maximum with respect to a single coordinate and a minimum in all other directions. Passage from one minimum to another one describes a chemical reaction and a saddle point between the two minima represents the transition state. Because of difficulties in representation of multidimensional hypersurfaces one-dimensional cross-sections through them are frequently used. The cross-sections may be compared to the potential energy curves of diatomic molecules and may often look similar to such curves. However, they must be interpreted with caution. For example, a saddle point may appear both as a minimum and maximum on two different cross-sections.

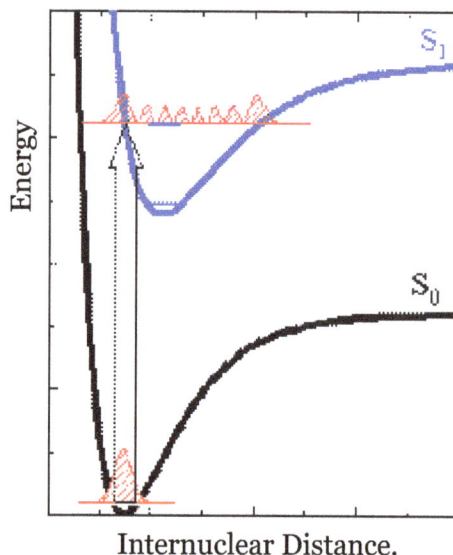

Internuclear Distance.

Franck-Condon Principle. The vibrational functions of two electronic states are approximately harmonic oscillator-like functions. The most probable position of the nuclei in the ground state corresponds to the maximum of the probability distribution function for the zero level (red curve). The energy gap between vibrational levels is usually large enough so that population of excited levels is small. An electronic transition caused by light absorption is represented by a vertical line (block arrow). The highest probability of the transition corresponds to the largest overlap between the ground-state and excited-state vibrational wavefunctions. The overlap is greatest for the S1 vibrational level whose classical turning point is near the equilibrium distance of the ground state.

Thermal reactions are generally considered to be adiabatic, i.e., they are represented by the motion on the lowest potential energy surface. Another way of putting this is that these reactions occur exclusively in the ground state. Therefore, knowledge of the ground-state potential surface is sufficient for modeling thermal reactivity with reaction rate theories. In contrast, the theoretical treatment of any photochemical reaction requires information about potential energy surfaces for more than one state. The photoreaction starts from the ground state of the reactant(s), necessarily proceeds via electronically excited state(s), and ends with the product(s) in the ground state. Therefore, photochemical reactions inevitably include diabatic processes, i.e., a transition from one potential surface to another. This statement should illuminate the complexity of the theoretical analysis of photoreactions, especially because reliable calculations of the potential energy surfaces for electronically excited states of reasonably large molecules still represent a challenge for computational chemistry. Nevertheless, many fundamental aspects of complex photoinduced reactions still can be understood from qualitative analysis of potential energy surfaces.

Upon light absorption, a molecular system may be transferred from the ground state to an electronically excited state. According to the Franck-Condon principle, this transition tends to occur between those vibrational levels of two electronic states that have the same nuclear configurations. The time required for the absorption of a light quantum (~1 fs) is much shorter than a characteristic time of a nuclear vibration (~100 fs), and therefore, the nuclei cannot change their relative positions during the act of excitation. In other words, transitions between two potential energy surfaces can be represented by vertical lines connecting them. In the course of a photochemical reaction there is a considerable time interval when the molecular system is out of the thermal equilibrium (a few ps in condensed phase, up to ms in low pressure gas phase reactions). It means that the population of vibrational energy levels may differ strongly from that predicted by the Boltzmann distribution. As a consequence of "vertical" electronic transitions and different equilibrium geometries of the ground and first excited state, immediately after excitation the molecular system will likely be in an excited vibrational state ("hot" molecule).

The amount of extra energy available for nuclear motion is a function of the excitation energy (wavelength). Vibrational excitation may also result from internal conversion or intersystem crossing, when electronic energy is converted into kinetic energy of the nuclei. It is known that internal conversion from S1 to S0 can be so fast in some systems that the thermal equilibration is first achieved only in the ground state. In solution, "hot" molecules in the first excited or ground state are quickly cooled down via interactions with the surroundings. Thermal equilibrium is normally established within a few picoseconds. Nevertheless this time is long enough to comprise several vibrational periods. The excess of kinetic energy may help the reactant(s) to overcome the barrier and relax into a new minimum. Chemical reactions of this type are called "hot". They preferentially occur in the gas phase at lower pressure where molecular collision frequency is much smaller than in the condensed phase.

Theoretical analysis of thermal reactions can be accomplished when minima and saddle points on the ground-state surface are allocated. The situation is much more complex for photochemical reactions. Difficulties emerge when one needs to explore several potential energy surfaces in detail. Luckily, only a few excited-state surfaces are of importance for the majority of photoreactions. Even so, topology of the three surfaces, S_0, S_1 and T_1, which are almost without exception needed to understand the photoreaction mechanism, may be extremely complex. Minima on S_1 and T_1 surfaces may be anticipated in the regions near the ground-state equilibrium geometries and near geometries, corresponding to intermolecular complexes. The latter minima reflect much larger polarizability of excited species and therefore higher affinity to other molecules. Excited complexes can be formed from two molecules of the same type (excimer), or from two different molecules (exciplex). Return from the minima of these two types to the ground state usually does not produce a chemical change unless significant geometrical changes accompany the excitation and/or multiple close-spaced minima exist on the ground state surface. Formaldehyde provides an example for large geometrical distortions in the excited state, the molecule is planar in S_0, and pyramidal in S_1 and T_1.

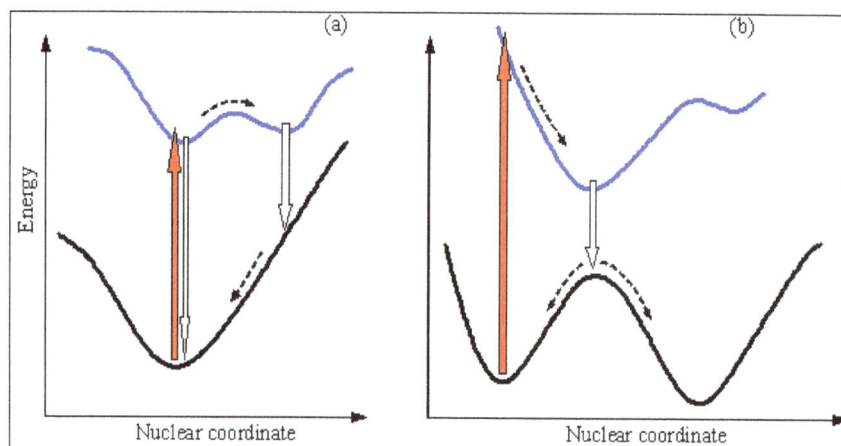

Schematic representation of the energy profiles corresponding to the ground state and the first excited state (a) for a system that undergoes an excited-state reaction but achieves no chemical conversion upon returning to the ground state and (b) for a system with partial conversion upon jumping to the ground state. Light absorption is represented by red block arrows, light emission by white block arrows.

In addition to localizing minima on the potential surfaces, finding the regions where the surfaces may cross or come very close to each other is of primary importance. The Born-Oppenheimer approximation is generally invalid in the vicinity of surface crossings and additional effects must be taken into account to describe the time evolution of the molecular species. The non-crossing rule states that potential energy curves can cross only if the electronic states have different symmetry (spatial or spin). Therefore the wavefunctions in the crossing region predicted by the simplest approximation has to be modified to avoid crossing of the potential energy curves. The non-crossing rule is strictly valid only for diatomic molecules. Intersection or touching of potential energy surfaces in polyatomic systems is generally allowed even if they belong to the states of the same symmetry. Recent studies showed that such crossing, also called conical intersection because of the topology of the surfaces at the crossing point, is quite common. The question whether a true conical intersection or avoided crossing is observed for a particular system of interest can be answered only with quantum mechanical calculations of high accuracy. These calculations recently became feasible for relatively large organic molecules, but reliable data are available just for a few systems.

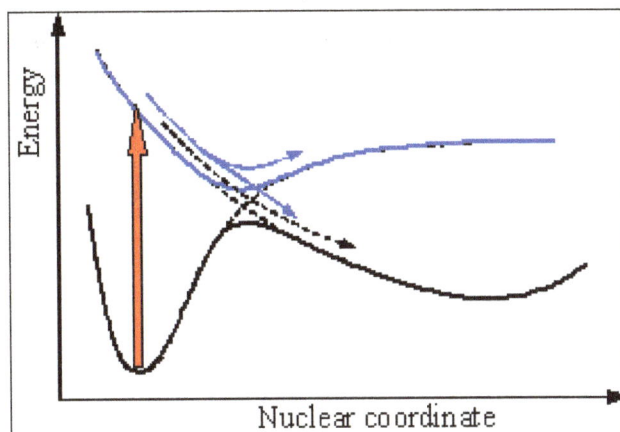

Adiabatic (solid) and non-adiabatic energy curves (dashed) for the S_0 and S_1 states. The light absorption is a vertical transition (block red arrow). Nuclear motion after excitation is governed by the S_1 curve. Blue arrows show the motion in the case of avoided crossing and the black broken arrow corresponds to the allowed crossing.

Two hypothetical surfaces for the ground- and an excited state are depicted in Figure. The fact that multidimensional potential-energy surfaces may have numerous regions where they come very close to each other is of great importance for understanding photochemical mechanisms. First, non-radiative transitions such as internal conversion and intersystem crossing have much higher probability in these regions. Second, conical intersections (or weakly avoided crossings) serve as bottlenecks through which the photoreaction passes on the way from excited-state species to the ground-state products. In this sense crossing points are analogous to the transition states on the adiabatic surfaces. An essential distinct feature of the conical intersection is the presence of two independent pathways for the reaction (path f) as compared to the single path through the saddle point.

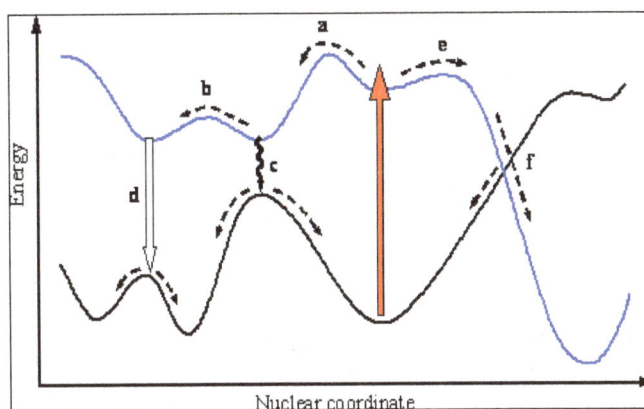

Potential energy surfaces of the ground and an excited state with various pathways (dashed lines) following the light absorption (red arrow).

"Vertical" excitation typically leads to vibrationally excited species. Thermal equilibrium may be established during the lifetime of the excited state, meaning that vibrational relaxation takes place and the photoreaction starting from a minimum on the excited-state surface is said to have an excited-state intermediate (path a). Return from the first or even the second minimum reached on the excited-state surface often does not produce a new species (right part of path c) and the whole sequence may be considered as a photophysical process. A typical example is the protolytic dissociation of 1-naphthol in the singlet excited state. The acidity of this molecule increases dramatically upon excitation (pKa = 9.2 and 0.4 for S_0 and S1) and proton is transferred to a suitable acceptor such as water. It has to be noted that figure does not account for all photoprocesses occurring in1-naphthol solutions.

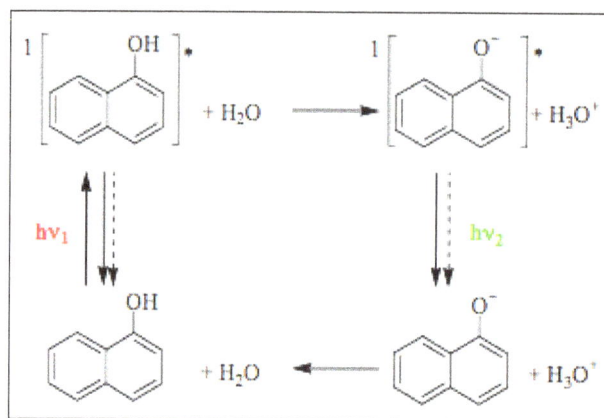

The primary excited-state intermediate in Figure may produce a new molecule in the excited state, which undergoes further modifications (path b), or returns to a new minimum on the ground-state surface (left part of path c). A jump from the excited-state surface can be accomplished via non-radiative transition (path c) or light emission (path d). An illustrative example of an excited-state intermediate in the photochemical reaction is the interaction of 9-cyanophenathrene with tetramethylethene in benzene that forms a cycloadduct via a singlet exciplex.

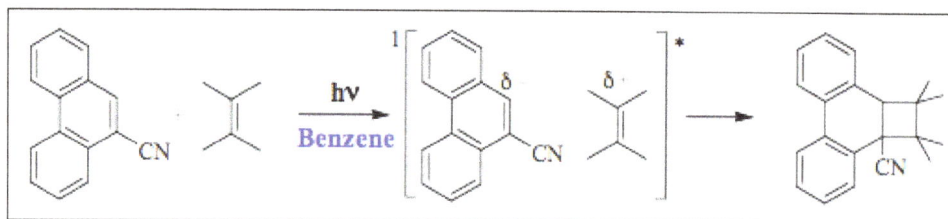

The reaction sequences represented by motion on the excited-state adiabatic surface are usually called adiabatic reactions. If the loss of excitation occurs anywhere on the reaction path between the points corresponding to reactants and products, then such photoreaction may be referred to as diabatic (also called non-adiabatic). It is also possible that the vibrational relaxation first occurs in the ground state. Such a photoreaction is called "direct". A direct reaction proceeds through a funnel (path f), which is a region of the potential energy surface where the probability for a jump from one energy surface to another one is very high. Funnels usually correspond to conical intersections or weakly avoided crossings. To characterize a molecule in a funnel one needs not only the positions of the nuclei but also their velocity vectors. In some systems passage through a conical intersection may also be separated from the excited-state minimum initially populated by a small barrier (paths a and c assuming that surfaces now cross at the point corresponding to path c). The presence of a S1-So conical intersection separated from the "vertical" geometry by a small barrier has been predicted for benzene. This funnel is responsible for the opening of efficient deactivation channel leading to disappearance of fluorescence and isomerization when the benzene molecule has enough vibrational energy to overcome the barrier.

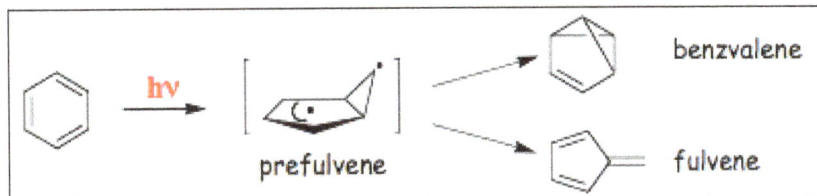

Factors Determining Outcome of a Photochemical Reaction

The wide variety of molecular mechanisms of photochemical reactions makes a general discussion of such factors very difficult. The chemical nature of the reactant(s) is definitely among the most important factors determining chemical reactivity initiated by light. However, a better understanding of this aspect may be gained from a closer examination of the individual groups of chemical compounds. The nature of excited states involved in a photoreaction is directly related to the electronic structure of the reactant(s).

Environmental variables, i.e., parameters that are not directly related to the chemical nature of the reacting systems, may also strongly affect photochemical reactivity. It is useful to distinguish between variables that are common for thermal and photochemical reactions, and those that are specific for the reactions of excited species. The first group includes reaction medium, reaction mixture composition, temperature, isotope effects to name the most important. The distinctive feature of photochemical reactions is that these parameters almost always operate under conditions when one or more photophysical processes compete with a photoreaction. The result of a photoinduced transformation can only be understood as the interplay of several processes corresponding to passages on and between at least two potential energy surfaces. We saw that even the simplest system, shown in figure, corresponds to parallel reactions in terms of reaction kinetics.

Reaction medium may directly modify the potential energy surfaces of the ground and excited states and hence affect the photoreactivity. The outcome of the two reactions presented in figures and changes dramatically when solvent polarity and hydrogen bonding capacity are changed. The protolytic photodissociation of 1-naphthol is completely suppressed in aprotic solvents because of unfavorable solvation energies both for the anion and proton. Under such conditions, proton transfer reaction cannot compete with the deactivation. The formation of two new products in the reaction of 9-cyanophenathrene with tetramethylethene is observed in methanol, because the exciplex dissociated into radical ions. It means that the potential energy minimum corresponding to the ion-radical pair shifts below that of the exciplex in polar solvents. The ion-radical formation is often followed by proton transfer reactions.

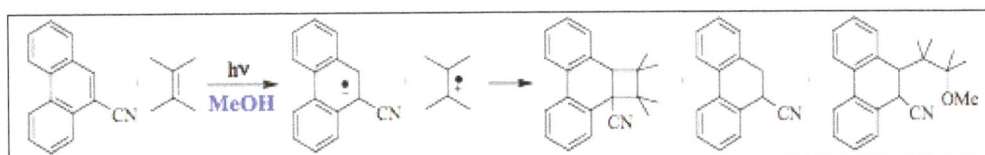

Solvent viscosity will strongly affect photoreactions where the encounter of two reactants or a substantial structural change are required. In highly viscous or solid solutions the loss of excitation via light emission or unimolecular non-radiative deactivation is more probable than a chemical modification of the excited species. On the other hand, slow diffusion in viscous solutions may prevent self-deactivation of the triplet state via a bimolecular process called triplet-triplet annihilation and enhance the efficiency of a photoreaction from this state. Triplet-triplet annihilation belongs to electronic-energy transfer processes, which may be classified as quenching of excited states. Quenching rate is a very important factor in discussing effects of medium and reaction mixture composition on photoreactivity. Quenching of excited states is a general phenomenon that is realized via different mechanisms. Any process that leads to the disappearance of the excited state of interest may be considered as quenching. In general it can be represented as:

$$S* + Q \xrightarrow{k_q} S' + Q$$

Notice that the quencher molecule Q may belong to the same kind of chemical species as the excited molecule, and be either in the ground or in an excited state. S' corresponds to the ground state or to an excited state of lower energy. we separated quenching described by figure from all other processes, including the photoreaction of interest introduced in figure. Obviously, this separation is just a matter of convention. Generally, any chemical reaction of the excited species can be considered as a quenching process for fluorescence. figure can easily be incorporated into the reaction scheme and into our kinetic analysis as an additional pseudo-unimolecular rate constant $k_q[Q]$.

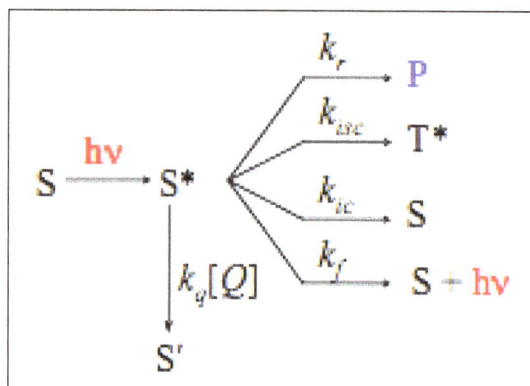

Kinetic scheme for a simple system with a photoreactive singlet state in the presence of a quencher.

In the presence of a quencher, Q, the observed lifetime of the excited molecule and therefore the quantum yield of the photoreaction may be significantly reduced.

$$\tau = \frac{1}{k_f + k_{ic} + k_{isc} + k_r + k_q[Q]}$$

$$\phi = k, \tau = \frac{k_r}{k_f + k_{ic} + k_{isc} + k_r + k_q[Q]}$$

Equation $\phi_0 = \dfrac{k_r[S*]}{W} = \dfrac{k_r}{k_f + k_{ic} + k_{isc} + k_r}$, $\tau_0 = \left(k_f + k_{ic} + k_{isc} + k_r\right)^{-1}$, $\tau = \dfrac{1}{k_f + k_{ic} + k_{isc} + k_r + k_q[Q]}$, and

$$\phi = k, \tau = \frac{k_r}{k_f + k_{ic} + k_{isc} + k_r + k_q[Q]} \text{ can be combined into a single one:}$$

$$\frac{\tau_0}{\tau} = \frac{\phi_0}{\phi} = 1 + k_q[Q]$$

where index "0" refers to the system without quenching. If we would consider the fluorescence quantum yield instead of the photoreaction yield, we would obtain a similar equation, which is known as the Stern-Volmer equation. The mechanism just considered corresponds to so-called dynamic quenching that results purely from encounters between excited molecules and the quencher. It is also conceivable that Q and S form a ground-state complex, which has a different reactivity and/or does not fluoresce. This situation is referred to as static quenching. In the case of static quenching, the quantum yield is diminished but the observed lifetime remains constant. In any event, the existence of quenching emphasizes the importance of the concentration as a controlling factor in photochemistry. For many systems, the quenching rate constant, k_q, is close to the diffusion-controlled limit which is of the order 10^{10} M-1s-1 at ambient temperature in liquid solutions. It means that quenching effects may become noticeable at the quencher concentrations > 1 mM and > 1 µM for the singlet state and triplet state with characteristic lifetimes of 10 ns and 10 µs, respectively. Thus even minor impurities may cause photoreaction quenching. The concentration of the photoreactive compound S may also play an important role if self-quenching takes place.

Because of the energy conservation law the excitation energy in a quenching process must be either dissipated in the form of thermal energy, or accumulated in the form of chemical energy of the quenching products or transferred to the quencher Q. According to these three possibilities one may distinguish physical mechanisms of quenching from chemical ones and from energy transfer. However, a clear cut is not always possible or worth making. The formation of excimers is frequently observed in solutions of aromatic hydrocarbons, such as anthracene or pyrene. The potential energy surfaces in these systems frequently look similar to that shown in Figure . Thus, the entire reaction sequence leads only to quenching of the excited monomer. The quenching will be seen in a reduced quantum yield of the monomer fluorescence and a monomer photoreaction. An illustrative example is 1-hydroxypyrene, which is a moderately strong photoacid in water. In the singlet excited states it readily transfers a proton to a suitable base such as acetate anion. But at higher concentrations of 1-hydroxypyrene, the quantum yield of the photoinduced proton transfer decreases because of the formation of the excimer, which is not as efficient as the proton donor. Exciplexes are typically more reactive, and provide examples for combined physical and chemical quenching.

Fluorescence self-quenching in aqueous solutions of dyes, such as fluorescein or eosin, has been known for more than 100 years. Several mechanisms involving collisional quenching, ground-state aggregation and energy transfer to the aggregates has been proposed to account for this phenomenon. In principle, quenching by the ground state could be observed for almost every excited species under conditions favoring the close proximity of two molecules. That is why it is often reported for systems with confined geometries such as those of surfactant assemblies. There exist many examples of self-quenching of the triplet state that plays a role in photochemistry. For example, the quenching of anthrone triplets by its ground state in benzene occurs with a rate constant close to 10^9 $M^{-1}S^{-1}$ and results in the formation of two radicals. The photoreactivity of 10, 10-dimethylanthrone differs dramatically, because methyl substituents prevent the reactive self-quenching.

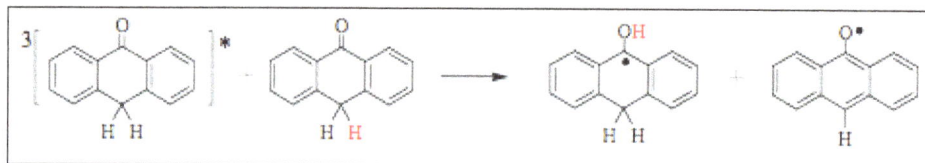

Compounds with heavy atoms and paramagnetic species increase the rate of intersystem crossing. It has to be emphasized that such molecules enhance the efficiency both of the S1 --> T1 and T1 --> So transitions, and should be considered as quenchers both for singlets and triplets. The yields of photochemical reactions originating from the singlet excited state, as a rule, are adversely affected by these quenchers. In contrast, the efficiency of photoconversion from the triplet state is usually increased because the triplet lifetime remains sufficiently long in the presence of a quencher, and the overall effects is largely determined by an increase in the yield of the triplets. An example is given by the photoreaction of anthracene with 1,3-cyclohexadiene which mainly forms product A. In the presence of methyl iodide (iodine is a heavy atom), the major product is compound B, which was also obtained in small quantities in the absence of the quencher. The results suggest that B is formed in a triplet-state reaction.

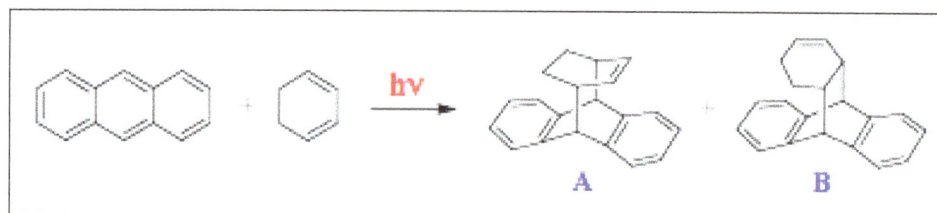

The most important paramagnetic species is molecular oxygen, which is known to be very efficient quencher of excited states. Quenching by O_2 is particularly important for the triplet state because of its long lifetime), so that even traces of oxygen may strongly affect photoreaction occurring through the triplet state. The ground state of O_2 is a triplet state. The first singlet excited state is only 22 kcal mol-1 above the ground state. This energy corresponds to near-IR radiation with a wavenumber of 7882.4 cm-1 or wavelength of 1269 nm. Singlet oxygen is a reactive species interacting with a wide variety of substrates. It can be generated using dyes with a high triplet yield, such as rose bengal or methylene blue. As mentioned above this process belongs to the type of quenching that is called electronic energy transfer.

The outcome of an energy-transfer process is the quenching of the luminescence or photoreaction associated with the donor and the initiation of the luminescence or photoreaction characteristic of the energy acceptor. The subsequent reactions of the acceptor are said to be sensitized. Electronic energy transfer can be described by figure where Q' has to be an excited-state species. Two general mechanisms of the energy transfer are distinguished: radiative and nonradiative. The radiative mechanism, often described as "trivial", is realized through the emission of light by the donor, and its absorption by the acceptor. Nonradiative energy transfer is a single-step process that requires the direct interaction of the donor and acceptor.

The specific variables of any photoreaction, as compared to thermal chemical processes, are the wavelength and intensity of excitation light. Wavelength dependence of the quantum yield or photoproduct composition may result from the occurrence of "hot" reactions or reactions from higher excited states (S_2, T_2, etc.). The latter processes have to be extremely fast to compete with internal conversion, which is typically accomplished within 1 ps. The presence of slowly inter-converting conformers or isomers with different absorption spectra may also cause wavelength-dependent photoreactions.

The intensity of excitation light is the key to multi-photon photochemistry. Because of the n-th power dependence of the absorption rate the photoreaction may become detectable only when the photon flux is above a certain threshold value. In one-photon photoreactions, primary processes are normally not affected by the light intensity. However, the overall reaction might be very sensitive to it, because relatively long-lived intermediates may come into play. These intermediates may absorb light and undergo photochemical reactions, or be involved in bimolecular reactions (e.g., triplet-triplet annihilation) that are strongly dependent on their concentration, and therefore light intensity.

2
Photochemical Reactions

A photochemical reaction takes place when a molecule comes in contact with the light. Luminescence, phosphorescence, fluorescence, photodegradation, photosensitization, photodissociation, etc. are some of the concepts that fall in this domain. The concepts elaborated in this chapter will help in gaining a better perspective about these photochemical reactions.

A photochemical reaction is a chemical reaction triggered when light energy is absorbed by a substance's molecules. This response leads the molecules to experience a temporary excited state, thus altering their physical and chemical properties from the substance's initial molecule.

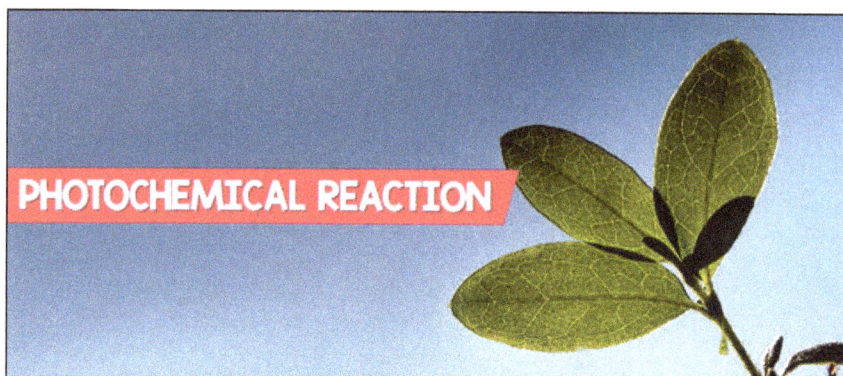

"The photochemical reaction is none other than a chemical reaction that starts with light being absorbed as a form of energy". Temporary peak states would be triggered while the molecules absorb light and there would be physical and chemical property differences to a large extent from the real molecules.

The resultant chemical structures could be separated, modified, mixed among the similar or different molecules along with the transfer of hydrogen atoms, electronic charge to separate molecules, protons, and electrons. The peak states in comparison to the real ground states are stronger reductants and acids that are stronger.

The mechanism of a photoreaction should ideally include a detailed characterization of the primary events as outlined by the classification of photochemical reaction pathways. The quantum yields and hence the rate constants of all relevant photophysical and photochemical processes, in addition to the information about the structure and fate of any reactive intermediates, their lifetimes and reactivities.

Photochemical Reaction Examples

The majority of processes on the other hand that we see in nature are photochemical ones. Our own ability to see the things in the world using the eyes is nothing but a photochemical reaction where a retinal that happens to be rhodopsin (photoreceptor cell molecule) changes its shape after sunlight or light absorption.

- Vitamin D which is required for bone and teeth development and even functioning of Kidney while helping skin growth is the chemical 7-dehydrocholesterol produced after exposure to sunlight.

- The ozone layer that is found in the earth's stratosphere is formed by the photochemical dissociation of molecular oxygen into oxygen atoms and these atoms reacting with molecules of oxygen to form ozone.

- The UltraViolet (UV) rays that are harmful to human DNA and skin cancer likes are caused by photochemical reactions.

- Different sorts of commercial processes and devices are heavily influenced by photochemical reactions and their peak states.

- Activities that we encounter in our daily lives like xerography, photography and so on are based on photochemical processes whereas complex activities like manufacturing of semiconductor chips, the printing of newspapers are done by the aid of UV rays.

- The examples explained above would have given you a fair idea about how chemical reactions like the photochemical ones have a major impact on our daily lives without which it would be impossible for life to sustain on our planet.

Luminescence

The term luminescence is used to describe a process by which light is produced other than by heating. The production of light from heat, or incandescence, is familiar to everyone. The Sun gives off both heat and light as a result of nuclear reactions in its core. An incandescent lightbulb gives off light when a wire filament inside the bulb is heated to white heat. One can read by the light of a candle flame because burning wax gives off both heat and light.

But light can also be produced by other processes in which heat is not involved. For example, fireflies produce light by means of chemical reactions that take place within their bodies. They convert a compound known as luciferin from one form into another. As that process occurs, light is given off.

Chemiluminescence

Chemiluminescence (also chemoluminescence) is the emission of light (luminescence), as the result of a chemical reaction. There may also be limited emission of heat. Given reactants A and B, with an excited intermediate ◊,

$$[A] + [B] \rightarrow [\lozenge] \rightarrow [Products] + \text{light}$$

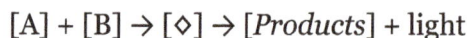

For example, if [A] is luminol and [B] is hydrogen peroxide in the presence of a suitable catalyst we have:

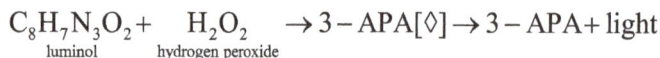

$$\underset{\text{luminol}}{C_8H_7N_3O_2} + \underset{\text{hydrogen peroxide}}{H_2O_2} \rightarrow 3-APA[\lozenge] \rightarrow 3-APA + \text{light}$$

where:

- 3-APA is 3-aminophthalate.

- 3-APA[$\lozenge$] is the vibronic excited state fluorescing as it decays to a lower energy level.

The decay of this excited state[$\lozenge$] to a lower energy level causes light emission. In theory, one photon of light should be given off for each molecule of reactant. This is equivalent to Avogadro's number of photons per mole of reactant. In actual practice, non-enzymatic reactions seldom exceed 1% Q_C, quantum efficiency.

A chemoluminescent reaction in an Erlenmeyer flask.

In a chemical reaction, reactants collide to form a transition state, the enthalpic maximum in a reaction coordinate diagram, which proceeds to the product. Normally, reactants form products of lesser chemical energy. The difference in energy between reactants and products, represented as, is turned into heat, physically realized as excitations in the vibrational state of the normal modes of the product. Since vibrational energy is generally much greater than the thermal agitation, it rapidly disperses in the solvent through molecular rotation. This is how exothermic reactions make their solutions hotter. In a chemiluminescent reaction, the direct product of the reaction is an excited electronic state. This state then decays into an electronic ground state and emits light through either an allowed transition (analogous to fluorescence) or a forbidden transition (analogous to phosphorescence), depending partly on the spin state of the electronic excited state formed.

Chemiluminescence differs from fluorescence or phosphorescence in that the electronic excited state is the product of a chemical reaction rather than of the absorption of a photon. It is the antithesis of a photochemical reaction, in which light is used to drive an endothermic chemical reaction. Here, light is generated from a chemically exothermic reaction. The chemiluminescence might be also induced by an electrochemical stimulus, in this case is called electrochemiluminescence.

Bioluminescence in nature: A male firefly mating
with a female of the species Lampyris noctiluca.

A standard example of chemiluminescence in the laboratory setting is the luminol test. Here, blood is indicated by luminescence upon contact with iron in hemoglobin. When chemiluminescence takes place in living organisms, the phenomenon is called bioluminescence. A light stick emits light by chemiluminescence.

Liquid-phase Reactions

Chemiluminescence in aqueous system is mainly caused by redox reactions.

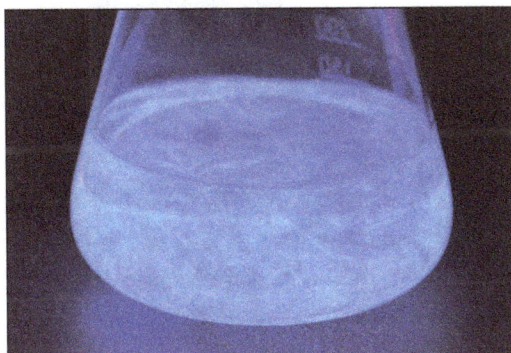

Chemiluminescence after a reaction of hydrogen peroxide and luminol.

- Luminol in an alkaline solution with hydrogen peroxide in the presence of iron or copper, or an auxiliary oxidant, produces chemiluminescence. The luminol reaction is

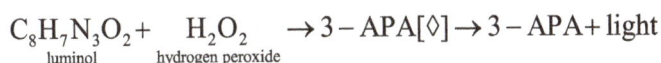

$$\underset{\text{luminol}}{C_8H_7N_3O_2} + \underset{\text{hydrogen peroxide}}{H_2O_2} \rightarrow 3 - APA[\lozenge] \rightarrow 3 - APA + \text{light}$$

Gas-phase Reactions

- One of the oldest known chemiluminescent reactions is that of elemental white phosphorus oxidizing in moist air, producing a green glow. This is a gas-phase reaction of phosphorus vapor, above the solid, with oxygen producing the excited states $(PO)_2$ and HPO.

- Another gas phase reaction is the basis of nitric oxide detection in commercial analytic instruments applied to environmental air-quality testing. Ozone is combined with nitric oxide to form nitrogen dioxide in an activated state.

$$NO + O_3 \rightarrow NO_2[\lozenge] + O_2$$

The activated $NO_2[\diamond]$ luminesces broadband visible to infrared light as it reverts to a lower energy state. A photomultiplier and associated electronics counts the photons that are proportional to the amount of NO present. To determine the amount of nitrogen dioxide, NO_2, in a sample (containing no NO) it must first be converted to nitric oxide, NO, by passing the sample through a converter before the above ozone activation reaction is applied. The ozone reaction produces a photon count proportional to NO that is proportional to NO_2 before it was converted to NO. In the case of a mixed sample that contains both NO and NO_2, the above reaction yields the amount of NO and NO_2 combined in the air sample, assuming that the sample is passed through the converter. If the mixed sample is not passed through the converter, the ozone reaction produces activated $NO_2[\diamond]$ only in proportion to the NO in the sample. The NO_2 in the sample is not activated by the ozone reaction. Though unactivated NO_2 is present with the activated $NO_2[\diamond]$, photons are emitted only by the activated species that is proportional to original NO. Final step: Subtract NO from (NO + NO_2) to yield NO_2.

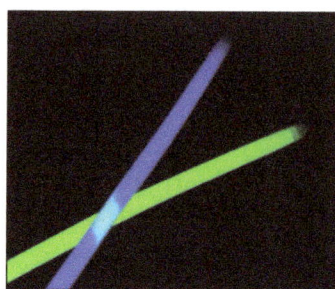

Green and blue glow sticks.

Infrared Chemiluminescence

In chemical kinetics, *infrared chemiluminiscence* (IRCL) refers to the emission of infrared photons from vibrationally excited product molecules immediately after their formation. The intensities of infrared emission lines from vibrationally excited molecules are used to measure the populations of vibrational states of product molecules.

The observation of IRCL was developed as a kinetic technique by John Polanyi, who used it to study the attractive or repulsive nature of the potential energy surface for gas-phase reactions. In general the IRCL is much more intense for reactions with an attractive surface, indicating that this type of surface leads to energy deposition in vibrational excitation. In contrast reactions with a repulsive potential energy surface lead to little IRCL, indicating that the energy is primarily deposited as translational energy.

Enhanced Chemiluminescence

Enhanced chemiluminescence is a common technique for a variety of detection assays in biology. A horseradish peroxidase enzyme (HRP) is tethered to an antibody that specifically recognizes the molecule of interest. This enzyme complex then catalyzes the conversion of the enhanced chemiluminescent substrate into a sensitized reagent in the vicinity of the molecule of interest, which on further oxidation by hydrogen peroxide, produces a triplet (excited) carbonyl, which emits light when it decays to the singlet carbonyl. Enhanced chemiluminescence allows detection of minute quantities of a biomolecule. Proteins can be detected down to femtomole quantities, well below the detection limit for most assay systems.

Applications

- Gas analysis: for determining small amounts of impurities or poisons in air. Other compounds can also be determined by this method (ozone, N-oxides, S-compounds). A typical example is NO determination with detection limits down to 1 ppb. Highly specialised chemiluminescence detectors have been used recently to determine concentrations as well as fluxes of NOx with detection limits as low as 5 ppt.

- Analysis of inorganic species in liquid phase.

- Analysis of organic species: useful with enzymes, where the substrate is not directly involved in the chemiluminescence reaction, but the product is.

- Detection and assay of biomolecules in systems such as ELISA and Western blots.

- DNA sequencing using pyrosequencing.

- Lighting objects: Chemiluminescence kites, emergency lighting, glow sticks (party decorations).

- Combustion analysis: Certain radical species (such as CH* and OH*) give off radiation at specific wavelengths. The heat release rate is calculated by measuring the amount of light radiated from a flame at those wavelengths.

- Children's toys.

- Glow sticks.

Biological Applications

Chemiluminescence has been applied by forensic scientists to solve crimes. In this case, they use luminol and hydrogen peroxide. The iron from the blood acts as a catalyst and reacts with the luminol and hydrogen peroxide to produce blue light for about 30 seconds. Because only a small amount of iron is required for chemiluminescence, trace amounts of blood are sufficient.

In biomedical research, the protein that gives fireflies their glow and its co-factor, luciferin, are used to produce red light through the consumption of ATP. This reaction is used in many applications, including the effectiveness of cancer drugs that choke off a tumor's blood supply. This form of bioluminescence imaging allows scientists to test drugs in the pre-clinical stages cheaply. Another protein, aequorin, found in certain jellyfish, produces blue light in the presence of calcium. It can be used in molecular biology to assess calcium levels in cells. What these biological reactions have in common is their use of adenosine triphosphate (ATP) as an energy source. Though the structure of the molecules that produce luminescence is different for each species, they are given the generic name of luciferin. Firefly luciferin can be oxidized to produce an excited complex. Once it falls back down to a ground state a photon is released. It is very similar to the reaction with luminol.

$$Luciferin + O_2 + ATP \xrightarrow{\text{Luciferase}} Oxyluciferin + CO_2 + AMP + PPi + light$$

Many organisms have evolved to produce light in a range of colors. At the molecular level, the difference in color arises from the degree of conjugation of the molecule, when an electron drops down from the excited state to the ground state. Deep sea organisms have evolved to produce light to lure and catch prey, as camouflage, or to attract others. Some bacteria even use bioluminescence to communicate. The common colors for the light emitted by these animals are blue and green because they have shorter wavelength than red and can transmit more easily in water.

Chemiluminescence is different from fluorescence. Hence the application of fluorescent proteins such as Green fluorescent protein is not a biological application of chemiluminescence.

Bioluminescence

Flying and glowing firefly, *Photinus pyralis.*

Bioluminescence is the production and emission of light by a living organism. It is a form of chemiluminescence. Bioluminescence occurs widely in marine vertebrates and invertebrates, as well as in some fungi, microorganisms including some bioluminescent bacteria and terrestrial invertebrates such as fireflies. In some animals, the light is bacteriogenic, produced by symbiotic organisms such as *Vibrio* bacteria; in others, it is autogenic, produced by the animals themselves.

In a general sense, the principal chemical reaction in bioluminescence involves some light-emitting molecule and an enzyme, generally called the luciferin and the luciferase, respectively. Because these are generic names, the luciferins and luciferases are often distinguished by including the species or group, i.e. Firefly luciferin. In all characterized cases, the enzyme catalyzes the oxidation of the luciferin.

Female Glowworm, *Lampyris noctiluca.*

Male and female of the species Lampyris noctiluca mating. The female of
this species is a larviform and has no wings, unlike the male.

In some species, the luciferase requires other cofactors, such as calcium or magnesium ions, and
sometimes also the energy-carrying molecule adenosine triphosphate (ATP). In evolution, luciferins vary little: one in particular, coelenterazine, is found in eleven different animal (phyla), though
in some of these, the animals obtain it through their diet. Conversely, luciferases vary widely between different species, and consequently bioluminescence has arisen over forty times in evolutionary history.

Both Aristotle and Pliny the Elder mentioned that damp wood sometimes gives off a glow and
many centuries later Robert Boyle showed that oxygen was involved in the process, both in
wood and in glow-worms. It was not until the late nineteenth century that bioluminescence
was properly investigated. The phenomenon is widely distributed among animal groups, especially in marine environments where dinoflagellates cause phosphorescence in the surface
layers of water. On land it occurs in fungi, bacteria and some groups of invertebrates, including insects.

The uses of bioluminescence by animals include counter-illumination camouflage, mimicry of other animals, for example to lure prey, and signalling to other individuals of the same species, such
as to attract mates. In the laboratory, luciferase-based systems are used in genetic engineering and
for biomedical research. Other researchers are investigating the possibility of using bioluminescent systems for street and decorative lighting, and a bioluminescent plant has been created.

Chemical Mechanism

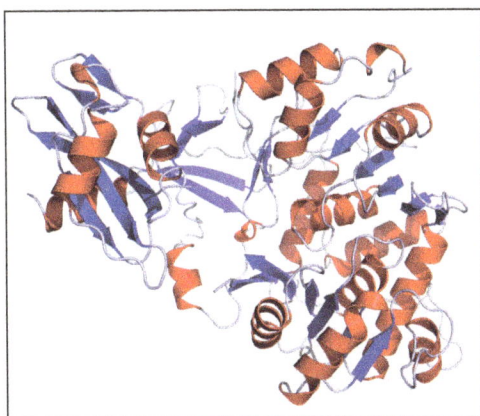

Protein structure of the luciferase of the firefly *Photinus pyralis*.
The enzyme is a much larger molecule than luciferin.

Bioluminescence is a form of chemiluminescence where light energy is released by a chemical reaction. This reaction involves a light-emitting pigment, the luciferin, and a luciferase, the enzyme component. Because of the diversity of luciferin/luciferase combinations, there are very few commonalities in the chemical mechanism. From currently studied systems, the only unifying mechanism is the role of molecular oxygen, though many examples have a concurrent release of carbon dioxide. For example, the firefly luciferin/luciferase reaction requires magnesium and ATP and produces carbon dioxide (CO_2), adenosine monophosphate (AMP) and pyrophosphate (PP) as waste products. Other cofactors may be required for the reaction, such as calcium (Ca^{2+}) for the photoprotein aequorin, or magnesium (Mg^{2+}) ions and ATP for the firefly luciferase. Generically, this reaction could be described as:

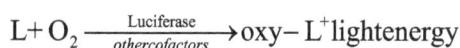

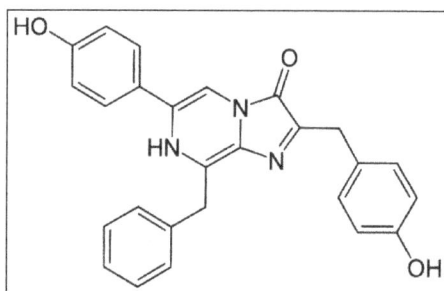

$$L + O_2 \xrightarrow[othercofactors]{Luciferase} oxy - L^+ lightenergy$$

Coelenterazine is a luciferin found in many different marine phyla from comb jellies to vertebrates. Like all luciferins, it is oxidised to produce light.

Instead of a luciferase, the jellyfish *Aequorea victoria* makes use of another type of protein called a photoprotein, in this case specifically aequorin. When calcium ions are added, the rapid catalysis creates a brief flash quite unlike the prolonged glow produced by luciferase. In a second, much slower, step luciferin is regenerated from the oxidised (oxyluciferin) form, allowing it to recombine with aequorin, in readiness for a subsequent flash. Photoproteins are thus enzymes, but with unusual reaction kinetics. Furthermore, some of the blue light released by aequorin in contact with calcium ions is absorbed by a green fluorescent protein, which in turn releases green light in a process called resonant energy transfer.

Overall, bioluminescence has arisen over forty times in evolutionary history. In evolution, luciferins tend to vary little: one in particular, coelenterazine, is the light emitting pigment for nine phyla (groups of very different organisms), including polycystine radiolaria, Cercozoa (Phaeodaria), protozoa, comb jellies, cnidaria including jellyfish and corals, crustaceans, molluscs, arrow worms and vertebrates (ray-finned fish). Not all these organisms synthesize coelenterazine: some of them obtain it through their diet. Conversely, luciferase enzymes vary widely and tend to be different in each species.

Distribution

Bioluminescence occurs widely among animals, especially in the open sea, including fish, jellyfish, comb jellies, crustaceans, and cephalopod molluscs; in some fungi and bacteria; and in various terrestrial invertebrates including insects. About 76% of the main taxa of deep-sea animals produce light. Most marine light-emission is in the blue and green light spectrum. However, some loose-jawed fish emit red and infrared light, and the genus *Tomopteris* emits yellow light.

Huge numbers of bioluminescent dinoflagellates
creating phosphorescence in breaking waves.

The most frequently encountered bioluminescent organisms may be the dinoflagellates present in the surface layers of the sea, which are responsible for the sparkling phosphorescence sometimes seen at night in disturbed water. At least eighteen genera exhibit luminosity. A different effect is the thousands of square miles of the ocean which shine with the light produced by bioluminescent bacteria, known as mareel or the milky seas effect.

Non-marine bioluminescence is less widely distributed, the two best-known cases being in fireflies and glow worms. Other invertebrates including insect larvae, annelids and arachnids possess bioluminescent abilities. Some forms of bioluminescence are brighter (or exist only) at night, following a circadian rhythm.

Uses in Nature

Bioluminescence has several functions in different taxa. Steven Haddock et al. (2010) list as more or less definite functions in marine organisms the following: defensive functions of startle, counterillumination (camouflage), misdirection (smoke screen), distractive body parts, burglar alarm and warning to deter settlers; offensive functions of lure, stun or confuse prey, illuminate prey, and mate attraction/recognition. It is much easier for researchers to detect that a species is able to produce light than to analyse the chemical mechanisms or to prove what function the light serves. In some cases the function is unknown, as with species in three families of earthworm (Oligochaeta), such as *Diplocardia longa* where the coelomic fluid produces light when the animal moves. The following functions are reasonably well established in the named organisms.

Counterillumination Camouflage

Principle of counterillumination camouflage in firefly squid, *Watasenia scintillans*. When seen from below by a predator, the bioluminescence helps to match the squid's brightness and colour to the sea surface above.

In many animals of the deep sea, including several squid species, bacterial bioluminescence is used for camouflage by counterillumination, in which the animal matches the overhead environmental light as seen from below. In these animals, photoreceptors control the illumination to match the brightness of the background. These light organs are usually separate from the tissue containing the bioluminescent bacteria. However, in one species, *Euprymna scolopes*, the bacteria are an integral component of the animal's light organ.

Attraction

A fungus gnat from New Zealand, *Arachnocampa luminosa*, lives in the predator-free environment of caves and its larvae emit bluish-green light. They dangle silken threads that glow and attract flying insects, and wind in their fishing-lines when prey becomes entangled. The bioluminescence of the larvae of another fungus gnat from North America which lives on streambanks and under overhangs has a similar function. *Orfelia fultoni* builds sticky little webs and emits light of a deep blue colour. It has an inbuilt biological clock and, even when kept in total darkness, turns its light on and off in a circadian rhythm.

Fireflies use light to attract mates. Two systems are involved according to species; in one, females emit light from their abdomens to attract males; in the other, flying males emit signals to which the sometimes sedentary females respond. Click beetles emit an orange light from the abdomen when flying and a green light from the thorax when they are disturbed or moving about on the ground. The former is probably a sexual attractant but the latter may be defensive. Larvae of the click beetle *Pyrophorus nyctophanus* live in the surface layers of termite mounds in Brazil. They light up the mounds by emitting a bright greenish glow which attracts the flying insects on which they feed.

In the marine environment, use of luminescence for mate attraction is chiefly known among ostracods, small shrimplike crustaceans, especially in the family Cyprididae. Pheromones may be used for long-distance communication, with bioluminescence used at close range to enable mates to "home in". A polychaete worm, the Bermuda fireworm creates a brief display, a few nights after the full moon, when the female lights up to attract males.

Defense

Many cephalopods, including at least 70 genera of squid, are bioluminescent. Some squid and small crustaceans use bioluminescent chemical mixtures or bacterial slurries in the same way as many squid use ink. A cloud of luminescent material is expelled, distracting or repelling a potential predator, while the animal escapes to safety. The deep sea squid *Octopoteuthis deletron* may autotomise portions of its arms which are luminous and continue to twitch and flash, thus distracting a predator while the animal flees.

Dinoflagellates may use bioluminescence for defence against predators. They shine when they detect a predator, possibly making the predator itself more vulnerable by attracting the attention of predators from higher trophic levels. Grazing copepods release any phytoplankton cells that flash, unharmed; if they were eaten they would make the copepods glow, attracting predators, so the phytoplankton's bioluminescence is defensive. The problem of shining stomach contents is solved (and the explanation corroborated) in predatory deep-sea fishes: their stomachs have a black lining able to keep the light from any bioluminescent fish prey which they have swallowed from attracting larger predators.

The sea-firefly is a small crustacean living in sediment. At rest it emits a dull glow but when disturbed it darts away leaving a cloud of shimmering blue light to confuse the predator. During World War II it was gathered and dried for use by the Japanese military as a source of light during clandestine operations.

The larvae of railroad worms (*Phrixothrix*) have paired photic organs on each body segment, able to glow with green light; these are thought to have a defensive purpose. They also have organs on the head which produce red light; they are the only terrestrial organisms to emit light of this colour.

Warning:

Aposematism is a widely used function of bioluminescence, providing a warning that the creature concerned is unpalatable. It is suggested that many firefly larvae glow to repel predators; some millipedes glow for the same purpose. Some marine organisms are believed to emit light for a similar reason. These include scale worms, jellyfish and brittle stars but further research is needed to fully establish the function of the luminescence. Such a mechanism would be of particular advantage to soft-bodied cnidarians if they were able to deter predation in this way. The limpet *Latia neritoides* is the only known freshwater gastropod that emits light. It produces greenish luminescent mucus which may have an anti-predator function. The marine snail *Hinea brasiliana* uses flashes of light, probably to deter predators. The blue-green light is emitted through the translucent shell, which functions as an efficient diffuser of light.

Communication

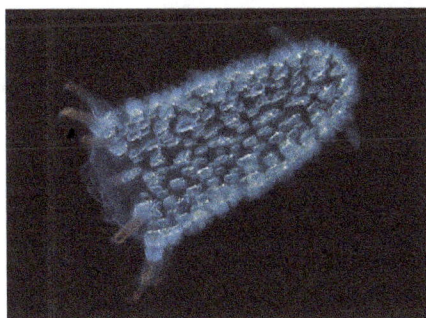

Pyrosoma, a colonial tunicate; each individual
zooid in the colony flashes a blue-green light.

Communication in the form of quorum sensing plays a role in the regulation of luminescence in many species of bacteria. Small extracellularly secreted molecules stimulate the bacteria to turn on genes for light production when cell density, measured by concentration of the secreted molecules, is high.

Pyrosomes are colonial tunicates and each zooid has a pair of luminescent organs on either side of the inlet siphon. When stimulated by light, these turn on and off, causing rhythmic flashing. No neural pathway runs between the zooids, but each responds to the light produced by other individuals, and even to light from other nearby colonies. Communication by light emission between the zooids enables coordination of colony effort, for example in swimming where each zooid provides part of the propulsive force.

Some bioluminous bacteria infect nematodes that parasitize Lepidoptera larvae. When these caterpillars die, their luminosity may attract predators to the dead insect thus assisting in the dispersal of both bacteria and nematodes. A similar reason may account for the many species of fungi that emit light. Species in the genera *Armillaria*, *Mycena*, *Omphalotus*, *Panellus*, *Pleurotus* and others do this, emitting usually greenish light from the mycelium, cap and gills. This may attract night-flying insects and aid in spore dispersal, but other functions may also be involved.

Quantula striata is the only known bioluminescent terrestrial mollusc. Pulses of light are emitted from a gland near the front of the foot and may have a communicative function, although the adaptive significance is not fully understood.

Mimicry

Bioluminescence is used by a variety of animals to mimic other species. Many species of deep sea fish such as the anglerfish and dragonfish make use of aggressive mimicry to attract prey. They have an appendage on their heads called an esca that contains bioluminescent bacteria able to produce a long-lasting glow which the fish can control. The glowing esca is dangled or waved about to lure small animals to within striking distance of the fish.

A deep sea anglerfish, *Bufoceratias wedli*, showing the esca (lure).

The cookiecutter shark uses bioluminescence to camouflage its underside by counterillumination, but a small patch near its pectoral fins remains dark, appearing as a small fish to large predatory fish like tuna and mackerel swimming beneath it. When such fish approach the lure, they are bitten by the shark.

Female *Photuris* fireflies sometimes mimic the light pattern of another firefly, *Photinus*, to attract its males as prey. In this way they obtain both food and the defensive chemicals named lucibufagins, which *Photuris* cannot synthesize.

South American giant cockroaches of the genus *Lucihormetica* were believed to be the first known example of defensive mimicry, emitting light in imitation of bioluminescent, poisonous click beetles. However, doubt has been cast on this assertion, and there is no conclusive evidence that the cockroaches are bioluminescent.

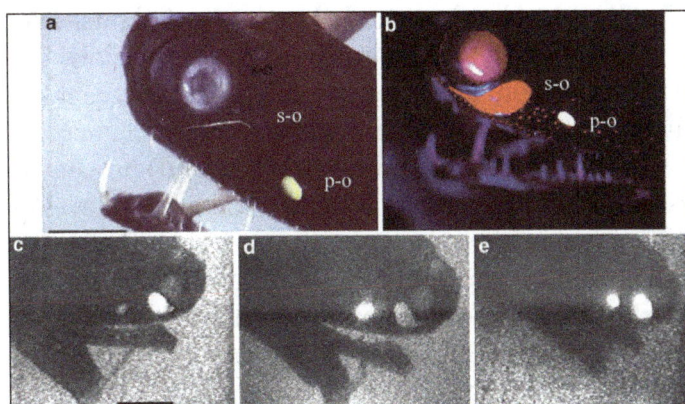

Flashing of photophores of black dragonfish, *Malacosteus niger*, showing red fluorescence.

Illumination

While most marine bioluminescence is green to blue, some deep sea barbeled dragonfishes in the genera *Aristostomias*, *Pachystomias* and *Malacosteus* emit a red glow. This adaptation allows the fish to see red-pigmented prey, which are normally invisible in the deep ocean environment where red light has been filtered out by the water column.

The black dragonfish (also called the northern stoplight loosejaw) *Malacosteus niger* is believed to be one of the only fish to produce a red glow. Its eyes, however, are insensitive to this wavelength; it has an additional retinal pigment which fluoresces blue-green when illuminated. This alerts the fish to the presence of its prey. The additional pigment is thought to be assimilated from chlorophyll derivatives found in the copepods which form part of its diet.

Biotechnology

Biology and Medicine

Bioluminescent organisms are a target for many areas of research. Luciferase systems are widely used in genetic engineering as reporter genes, each producing a different colour by fluorescence, and for biomedical research using bioluminescence imaging. For example, the firefly luciferase gene was used as early as 1986 for research using transgenic tobacco plants. *Vibrio* bacteria symbiose with marine invertebrates such as the Hawaiian bobtail squid (*Euprymna scolopes*), are key experimental models for bioluminescence. Bioluminescent activated destruction is an experimental cancer treatment. The optogenetics which involves the use of light to control cells in living tissue, typically neurons, that have been genetically modified to express light-sensitive ion channels, and also see biophoton, a photon of non-thermal origin in the visible and ultraviolet spectrum emitted from a biological system.

Light Production

The structures of photophores, the light producing organs in bioluminescent organisms, are being investigated by industrial designers. Engineered bioluminescence could perhaps one day be used to reduce the need for street lighting, or for decorative purposes if it becomes possible to produce light that is both bright enough and can be sustained for long periods at a workable price. The gene that makes the tails of fireflies glow has been added to mustard plants. The plants glow faintly for an hour when touched, but a sensitive camera is needed to see the glow. University of Wisconsin–Madison is researching the use of genetically engineered bioluminescent E. coli bacteria, for use as bioluminescent bacteria in a light bulb. In 2011, Philips launched a microbial system for ambience lighting in the home. An iGEM team from Cambridge (England) has started to address the problem that luciferin is consumed in the light-producing reaction by developing a genetic biotechnology part that codes for a luciferin regenerating enzyme from the North American firefly; this enzyme "helps to strengthen and sustain light output". In 2016, Glowee, a French company started selling bioluminescent lights, targeting shop fronts and municipal street signs as their main markets. France has a law that forbids retailers and offices from illuminating their windows between 1 and 7 in the morning in order to minimise energy consumption and pollution. Glowee hoped their product would get around this ban. They used bacteria called *Aliivibrio fischeri* which glow in the dark, but the maximum lifetime of their product was three days.

Photoluminescence

Photoluminescence is light emission from any form of matter after the absorption of photons (electromagnetic radiation). It is one of many forms of luminescence (light emission) and is initiated by photoexcitation (i.e. photons that excite electrons to a higher energy level in an atom), hence the prefix *photo-*. Following excitation various relaxation processes typically occur in which other photons are re-radiated. Time periods between absorption and emission may vary: ranging from short femtosecond-regime for emission involving free-carrier plasma in inorganic semiconductors up to milliseconds for Phosphorescence processes in molecular systems; and under special circumstances delay of emission may even span to minutes or hours.

Fluorescent solutions under UV-light. Absorbed photons are rapidly re-emitted under longer electromagnetic wavelengths.

Observation of photoluminescence at a certain energy can be viewed as an indication that an electron populated an excited state associated with this transition energy.

While this is generally true in atoms and similar systems, correlations and other more complex phenomena also act as sources for photoluminescence in many-body systems such as semiconductors. A theoretical approach to handle this is given by the semiconductor luminescence equations.

Forms

Photoluminescence processes can be classified by various parameters such as the energy of the exciting photon with respect to the emission. Resonant excitation describes a situation in which photons of a particular wavelength are absorbed and equivalent photons are very rapidly re-emitted. This is often referred to as resonance fluorescence. For materials in solution or in the gas phase, this process involves electrons but no significant internal energy transitions involving molecular features of the chemical substance between absorption and emission. In crystalline inorganic semiconductors where an electronic band structure is formed, secondary emission can be more complicated as events may contain both coherent contributions such as resonant Rayleigh scattering where a fixed phase relation with the driving light field is maintained (i.e. energetically elastic processes where no losses are involved), and incoherent contributions (or inelastic modes where some energy channels into an auxiliary loss mode).

The latter originate, e.g., from the radiative recombination of excitons, Coulomb-bound electron-hole pair states in solids. Resonance fluorescence may also show significant quantum optical correlations.

More processes may occur when a substance undergoes internal energy transitions before re-emitting the energy from the absorption event. Electrons change energy states by either resonantly gaining energy from absorption of a photon or losing energy by emitting photons. In chemistry-related disciplines, one often distinguishes between fluorescence and phosphorescence. The former is typically a fast process, yet some amount of the original energy is dissipated so that re-emitted light photons will have lower energy than did the absorbed excitation photons. The re-emitted photon in this case is said to be red shifted, referring to the reduced energy it carries following this loss. For phosphorescence, electrons which absorbed photons, undergo intersystem crossing where they enter into a state with altered spin multiplicity, usually a triplet state. Once the excited electron is transferred into this triplet state, electron transition (relaxation) back to the lower singlet state energies is quantum mechanically forbidden, meaning that it happens much more slowly than other transitions. The result is a slow process of radiative transition back to the singlet state, sometimes lasting minutes or hours. This is the basis for "glow in the dark" substances.

Photoluminescence is an important technique for measuring the purity and crystalline quality of semiconductors such as GaN and InP and for quantification of the amount of disorder present in a system.

Time-resolved photoluminescence (TRPL) is a method where the sample is excited with a light pulse and then the decay in photoluminescence with respect to time is measured. This technique is useful for measuring the minority carrier lifetime of III-V semiconductors like gallium arsenide (GaAs).

Photoluminescence Properties of Direct-gap Semiconductors

In a typical PL experiment, a semiconductor is excited with a light-source that provides photons with an energy larger than the bandgap energy. The incoming light excites a polarization that can be described with the semiconductor Bloch equations. Once the photons are absorbed, electrons and holes are formed with finite momenta in the conduction and valence bands, respectively. The excitations then undergo energy and momentum relaxation towards the band gap minimum. Typical mechanisms are Coulomb scattering and the interaction with phonons. Finally, the electrons recombine with holes under emission of photons.

Ideal, defect-free semiconductors are many-body systems where the interactions of charge-carriers and lattice vibrations have to be considered in addition to the light-matter coupling. In general, the PL properties are also extremely sensitive to internal electric fields and to the dielectric environment (such as in photonic crystals) which impose further degrees of complexity. A precise microscopic description is provided by the semiconductor luminescence equations.

Ideal Quantum-well Structures

An ideal, defect-free semiconductor quantum well structure is a useful model system to illustrate the fundamental processes in typical PL experiments.

The fictive model structure has two confined quantized electronic and two hole subbands, e_1, e_2 and h_1, h_2, respectively. The linear absorption spectrum of such a structure shows the exciton resonances of the first (e1h1) and the second quantum well subbands (e_2, h_2), as well as the absorption from the corresponding continuum states and from the barrier.

Photoexcitation

In general, three different excitation conditions are distinguished: resonant, quasi-resonant, and non-resonant. For the resonant excitation, the central energy of the laser corresponds to the lowest exciton resonance of the quantum well. No or only a negligible amount of the excess energy is injected to the carrier system. For these conditions, coherent processes contribute significantly to the spontaneous emission. The decay of polarization creates excitons directly. The detection of PL is challenging for resonant excitation as it is difficult to discriminate contributions from the excitation, i.e., stray-light and diffuse scattering from surface roughness. Thus, speckle and resonant Rayleigh-scattering are always superimposed to the incoherent emission.

In case of the non-resonant excitation, the structure is excited with some excess energy. This is the typical situation used in most PL experiments as the excitation energy can be discriminated using a spectrometer or an optical filter. One has to distinguish between quasi-resonant excitation and barrier excitation.

For quasi-resonant conditions, the energy of the excitation is tuned above the ground state but still below the barrier absorption edge, for example, into the continuum of the first subband. The polarization decay for these conditions is much faster than for resonant excitation and coherent contributions to the quantum well emission are negligible. The initial temperature of the carrier system is significantly higher than the lattice temperature due to the surplus energy of the injected carriers. Finally, only the electron-hole plasma is initially created. It is then followed by the formation of excitons.

In case of barrier excitation, the initial carrier distribution in the quantum well strongly depends on the carrier scattering between barrier and the well.

Relaxation

Initially, the laser light induces coherent polarization in the sample, i.e., the transitions between electron and hole states oscillate with the laser frequency and a fixed phase. The polarization dephases typically on a sub-100 fs time-scale in case of nonresonant excitation due to ultra-fast Coulomb- and phonon-scattering.

The dephasing of the polarization leads to creation of populations of electrons and holes in the conduction and the valence bands, respectively. The lifetime of the carrier populations is rather long, limited by radiative and non-radiative recombination such as Auger recombination. During this lifetime a fraction of electrons and holes may form excitons. The formation rate depends on the experimental conditions such as lattice temperature, excitation density, as well as on the general material parameters, e.g., the strength of the Coulomb-interaction or the exciton binding energy.

The characteristic time-scales are in the range of hundreds of picoseconds in GaAs; they appear to be much shorter in wide-gap semiconductors.

Directly after the excitation with short (femtosecond) pulses and the quasi-instantaneous decay of the polarization, the carrier distribution is mainly determined by the spectral width of the excitation, e.g., a laser pulse. The distribution is thus highly non-thermal and resembles a Gaussian distribution, centered at a finite momentum. In the first hundreds of femtoseconds, the carriers are scattered by phonons, or at elevated carrier densities via Coulomb-interaction. The carrier system successively relaxes to the Fermi–Dirac distribution typically within the first picosecond. Finally, the carrier system cools down under the emission of phonons. This can take up to several nanoseconds, depending on the material system, the lattice temperature, and the excitation conditions such as the surplus energy.

Initially, the carrier temperature decreases fast via emission of optical phonons. This is quite efficient due to the comparatively large energy associated with optical phonons, (36meV or 420K in GaAs) and their rather flat dispersion, allowing for a wide range of scattering processes under conservation of energy and momentum. Once the carrier temperature decreases below the value corresponding to the optical phonon energy, acoustic phonons dominate the relaxation. Here, cooling is less efficient due their dispersion and small energies and the temperature decreases much slower beyond the first tens of picoseconds. At elevated excitation densities, the carrier cooling is further inhibited by the so-called hot-phonon effect. The relaxation of a large number of hot carriers leads to a high generation rate of optical phonons which exceeds the decay rate into acoustic phonons. This creates a non-equilibrium "over-population" of optical phonons and thus causes their increased reabsorption by the charge-carriers significantly suppressing any cooling. Thus, a system cools slower, the higher the carrier density is.

Radiative Recombination

The emission directly after the excitation is spectrally very broad, yet still centered in the vicinity of the strongest exciton resonance. As the carrier distribution relaxes and cools, the width of the PL peak decreases and the emission energy shifts to match the ground state of the exciton (such as an electron) for ideal samples without disorder. The PL spectrum approaches its quasi-steady-state shape defined by the distribution of electrons and holes. Increasing the excitation density will change the emission spectra. They are dominated by the excitonic ground state for low densities. Additional peaks from higher subband transitions appear as the carrier density or lattice temperature are increased as these states get more and more populated. Also, the width of the main PL peak increases significantly with rising excitation due to excitation-induced dephasing and the emission peak experiences a small shift in energy due to the Coulomb-renormalization and phase-filling.

In general, both exciton populations and plasma, uncorrelated electrons and holes, can act as sources for photoluminescence as described in the semiconductor-luminescence equations. Both yield very similar spectral features which are difficult to distinguish; their emission dynamics, however, vary significantly. The decay of excitons yields a single-exponential decay function since the probability of their radiative recombination does not depend on the carrier density. The probability of spontaneous emission for uncorrelated electrons and holes, is approximately proportional to the product of electron and hole populations eventually leading to a non-single-exponential decay described by a hyperbolic function.

Effects of Disorder

Real material systems always incorporate disorder. Examples are structural defects in the lattice or disorder due to variations of the chemical composition. Their treatment is extremely challenging for microscopic theories due to the lack of detailed knowledge about perturbations of the ideal structure. Thus, the influence of the extrinsic effects on the PL is usually addressed phenomenologically. In experiments, disorder can lead to localization of carriers and hence drastically increase the photoluminescence life times as localized carriers cannot as easily find nonradiative recombination centers as can free ones.

Researchers from the King Abdullah University of Science and Technology (KAUST) have studied the photoinduced entropy (i.e. thermodynamic disorder) of InGaN/GaN p-i-n double-heterostructure and AlGaN nanowires using temperature-dependent photoluminescence. They defined the photoinduced entropy as a thermodynamic quantity that represents the unavailability of a system's energy for conversion into useful work due to carrier recombination and photon emission. They have also related the change in entropy generation to the change in photocarrier dynamics in the nanowire active regions using results from time-resolved photoluminescence study. They hypothesized that the amount of generated disorder in the InGaN layers eventually increases as the temperature approaches room temperature because of the thermal activation of surface states, while an insignificant increase was observed in AlGaN nanowires, indicating lower degrees of disorder-induced uncertainty in the wider bandgap semiconductor. To study the photoinduced entropy, the scientists have developed a mathematical model that considers the net energy exchange resulting from photoexcitation and photoluminescence.

Photoluminescent Material for Temperature Detection

In phosphor thermometry, the temperature dependence of the photoluminescence process is exploited to measure temperature.

Experimental Methods

Photoluminescence spectroscopy is a widely used technique for characterisation of the optical and electronic properties of semiconductors and molecules. In chemistry, it is more often referred to as fluorescence spectroscopy, but the instrumentation is the same. The relaxation processes can be studied using Time-resolved fluorescence spectroscopy to find the decay lifetime of the photoluminescence. These techniques can be combined with microscopy, to map the intensity (Confocal microscopy) or the lifetime (Fluorescence-lifetime imaging microscopy) of the photoluminescence across a sample (e.g. a semiconducting wafer, or a biological sample that has been marked with fluorescent molecules).

Phosphorescence

Phosphorescence is a type of photoluminescence related to fluorescence. Unlike fluorescence, a phosphorescent material does not immediately re-emit the radiation it absorbs. The slower time scales of the re-emission are associated with "forbidden" energy state transitions in quantum mechanics. As these transitions occur very slowly in certain materials, absorbed radiation is re-emitted at a lower intensity for up to several hours after the original excitation.

Everyday examples of phosphorescent materials are the glow-in-the-dark toys, stickers, paint, wristwatch and clock dials that glow after being charged with a bright light such as in any normal reading or room light. Typically, the glow slowly fades out, sometimes within a few minutes or up to a few hours in a dark room.

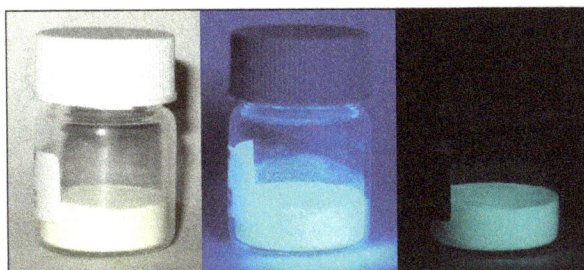

Phosphorescent, europium-doped strontium silicate-aluminate oxide powder under visible light, long-wave UV light, and in total darkness.

Around 1604, Vincenzo Casciarolo discovered a "lapis solaris" near Bologna, Italy. Once heated in an oxygen-rich furnace, it thereafter absorbed sunlight and glowed in the dark. The study of phosphorescent materials led to the discovery of radioactivity in 1896.

Explanation of Phosphorescence

Simple

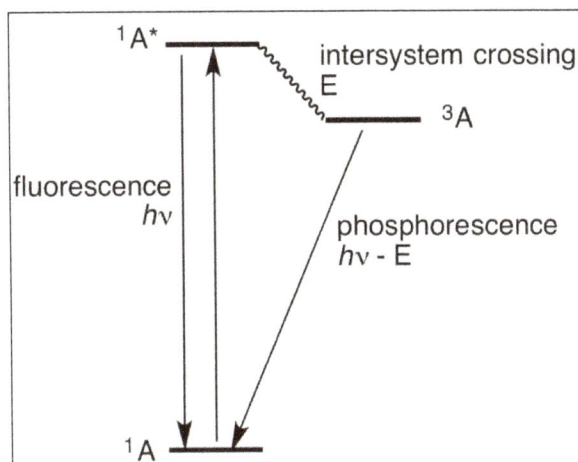

Jablonski diagram of an energy scheme used to explain the difference between fluorescence and phosphorescence. The excitation of molecule A to its singlet excited state ($^1A^*$) is followed by intersystem crossing to the triplet state (^3A) that relaxes to the ground state by phosphorescence.

In simple terms, phosphorescence is a process in which energy absorbed by a substance is released relatively slowly in the form of light. This is in some cases the mechanism used for "glow-in-the-dark" materials which are "charged" by exposure to light. Unlike the relatively swift reactions in fluorescence, such as those seen in a common fluorescent tube, phosphorescent materials "store" absorbed energy for a longer time, as the processes required to re-emit energy occur less often.

Quantum Mechanical

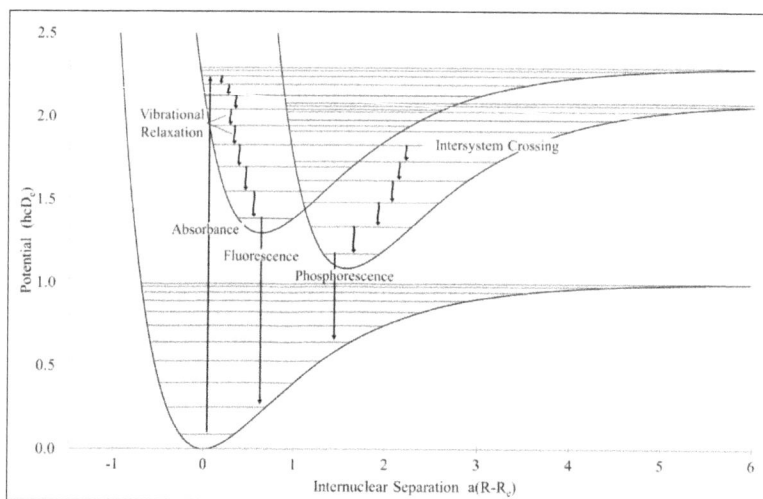

After an electron absorbs a photon of high energy, it may undergo vibrational relaxations and intersystem crossing to another spin state. Again the system relaxes vibrationally in the new spin state and eventually emits light by phosphorescence.

Most photoluminescent events, in which a chemical substrate absorbs and then re-emits a photon of light, are fast, in the order of 10 nanoseconds. Light is absorbed and emitted at these fast time scales in cases where the energy of the photons involved matches the available energy states and allowed transitions of the substrate. In the special case of phosphorescence, the electron which absorbed the photon (energy) undergoes an unusual intersystem crossing into an energy state of higher *spin multiplicity*, usually a triplet state. As a result, the excited electron can become trapped in the triplet state with only "forbidden" transitions available to return to the lower energy singlet state. These transitions, although "forbidden", will still occur in quantum mechanics but are kinetically unfavored and thus progress at significantly slower time scales. Most phosphorescent compounds are still relatively fast emitters, with triplet lifetimes on the order of milliseconds. However, some compounds have triplet lifetimes up to minutes or even hours, allowing these substances to effectively store light energy in the form of very slowly degrading excited electron states. If the phosphorescent quantum yield is high, these substances will release significant amounts of light over long time scales, creating so-called "glow-in-the-dark" materials.

Equation

$$S_0 + hv \rightarrow S_1 \rightarrow T_1 \rightarrow S_0 + hv'$$

where S is a singlet and T a triplet whose subscripts denote states (0 is the ground state, and 1 the excited state). Transitions can also occur to higher energy levels, but the first excited state is denoted for simplicity.

Materials

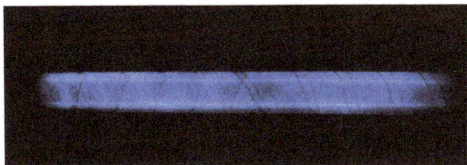

An extremely intense pulse of UV light in a flashtube produced
this blue phosphorescence in the fused silica envelope.

Common pigments used in phosphorescent materials include zinc sulfide and strontium aluminate. Use of zinc sulfide for safety related products dates back to the 1930s. However, the development of strontium aluminate, with a luminance approximately 10 times greater than zinc sulfide,
has relegated most zinc sulfide based products to the novelty category. Strontium aluminate based
pigments are now used in exit signs, pathway marking, and other safety related signage.

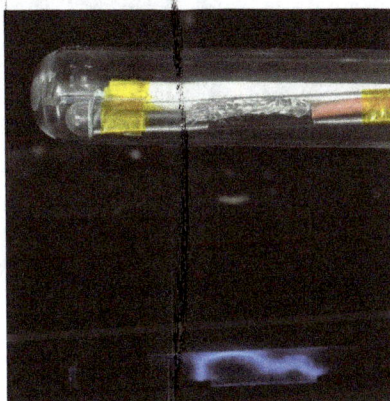

Phosphorescence of the quartz ignition tube of an air-gap flash.

- Phosphorescent pigments – zinc sulfide vs. strontium aluminate

Left: zinc sulfide Right: strontium
aluminate.

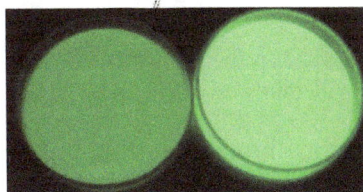

Pigments in the
dark.

Pigments in the dark
after 4 min.

- Phosphorescent

Phosphorescent pigment
red (calcium sulfide).

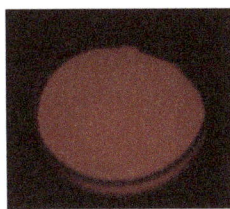

Phosphorescent pig-
ment red in the dark.

Phosphorescent pigment blue
(alkaline earth metal silicate).

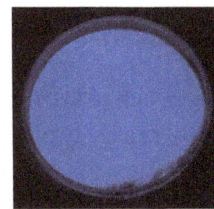

Phosphorescent pigment
blue in the dark.

Fluorescence

Fluorescence is the emission of light by a substance that has absorbed light or other electromagnetic radiation. It is a form of luminescence. In most cases, the emitted light has a longer wavelength, and therefore lower energy, than the absorbed radiation. The most striking example of fluorescence occurs when the absorbed radiation is in the ultraviolet region of the spectrum, and thus invisible to the human eye, while the emitted light is in the visible region, which gives the fluorescent substance a distinct color that can be seen only when exposed to UV light. Fluorescent materials cease to glow nearly immediately when the radiation source stops, unlike phosphorescent materials, which continue to emit light for some time after.

Fluorescent minerals emit visible light when exposed to ultraviolet light.

Willemite and calcite in UV light.

Fluorescence has many practical applications, including mineralogy, gemology, medicine, chemical sensors (fluorescence spectroscopy), fluorescent labelling, dyes, biological detectors, cosmic-ray detection, and, most commonly, fluorescent lamps. Fluorescence also occurs frequently in nature in some minerals and in various biological states in many branches of the animal kingdom.

Physical Principles

Photochemistry

Fluorescence occurs when an orbital electron of a molecule, atom, or nanostructure, relaxes to its ground state by emitting a photon from an excited singlet state:

- Excitation: $S_0 + h\nu_{ex} \rightarrow S_1$

- Fluorescence (emission): $S_1 \rightarrow S_0 + h\nu_{em} + \text{heat}$

Here is a generic term for photon energy with h = Planck's constant and v = frequency of light. The specific frequencies of exciting and emitted lights are dependent on the particular system.

S_0 is called the ground state of the fluorophore (fluorescent molecule), and S_1 is its first (electronically) excited singlet state.

A molecule in S_1 can relax by various competing pathways. It can undergo *non-radiative* relaxation in which the excitation energy is dissipated as heat (vibrations) to the solvent. Excited organic molecules can also relax via conversion to a triplet state, which may subsequently relax via phosphorescence, or by a secondary non-radiative relaxation step.

Relaxation from S_1 can also occur through interaction with a second molecule through fluorescence quenching. Molecular oxygen (O_2) is an extremely efficient quencher of fluorescence just because of its unusual triplet ground state.

In most cases, the emitted light has a longer wavelength, and therefore lower energy, than the absorbed radiation; this phenomenon is known as the Stokes shift. However, when the absorbed electromagnetic radiation is intense, it is possible for one electron to absorb two photons; this two-photon absorption can lead to emission of radiation having a shorter wavelength than the absorbed radiation. The emitted radiation may also be of the same wavelength as the absorbed radiation, termed "resonance fluorescence".

Molecules that are excited through light absorption or via a different process (e.g. as the product of a reaction) can transfer energy to a second 'sensitized' molecule, which is converted to its excited state and can then fluoresce.

Quantum Yield

The fluorescence quantum yield gives the efficiency of the fluorescence process. It is defined as the ratio of the number of photons emitted to the number of photons absorbed.

$$\Phi = \frac{\text{Number of photons emitted}}{\text{Number of photons absorbed}}$$

The maximum possible fluorescence quantum yield is 1.0 (100%); each photon absorbed results in a photon emitted. Compounds with quantum yields of 0.10 are still considered quite fluorescent. Another way to define the quantum yield of fluorescence is by the rate of excited state decay:

$$\Phi = \frac{k_f}{\sum_i k_i}$$

where k_f is the rate constant of spontaneous emission of radiation and,

$$\sum_i k_i$$

is the sum of all rates of excited state decay. Other rates of excited state decay are caused by mechanisms other than photon emission and are, therefore, often called "non-radiative rates", which can include: dynamic collisional quenching, near-field dipole-dipole interaction (or resonance energy transfer), internal conversion, and intersystem crossing. Thus, if the rate of any pathway changes, both the excited state lifetime and the fluorescence quantum yield will be affected.

Fluorescence quantum yields are measured by comparison to a standard. The quinine salt *quinine sulfate* in a sulfuric acid solution is a common fluorescence standard.

Lifetime

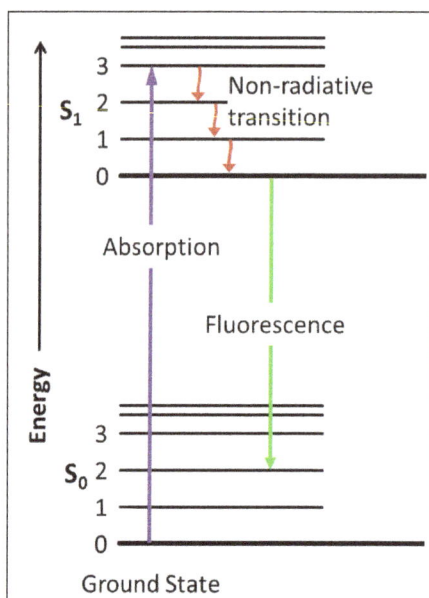

Jablonski diagram. After an electron absorbs a high-energy photon the system is excited electronically and vibrationally. The system relaxes vibrationally, and eventually fluoresces at a longer wavelength.

The fluorescence lifetime refers to the average time the molecule stays in its excited state before emitting a photon. Fluorescence typically follows first-order kinetics:

$$[S_1] = [S_1]_0 \, e^{-rt}$$

where $[S_1]$ is the concentration of excited state molecules at time t, $[S_1]_0$ is the initial concentration and Γ is the decay rate or the inverse of the fluorescence lifetime. This is an instance of exponential decay. Various radiative and non-radiative processes can de-populate the excited state. In such case the total decay rate is the sum over all rates:

$$\Gamma_{tot} = \Gamma_{rad} + \Gamma_{nrad}$$

where Γ_{tot} is the total decay rate, Γ_{rad} the radiative decay rate and Γ_{nrad} the non-radiative decay rate. It is similar to a first-order chemical reaction in which the first-order rate constant is the sum of all of the rates (a parallel kinetic model). If the rate of spontaneous emission, or any of the other rates are fast, the lifetime is short. For commonly used fluorescent compounds, typical excited state decay times for photon emissions with energies from the UV to near infrared are within the range of 0.5 to 20 nanoseconds. The fluorescence lifetime is an important parameter for practical applications of fluorescence such as fluorescence resonance energy transfer and fluorescence-life-time imaging microscopy.

Jablonski Diagram

The Jablonski diagram describes most of the relaxation mechanisms for excited state molecules. The diagram alongside shows how fluorescence occurs due to the relaxation of certain excited electrons of a molecule.

Fluorescence Anisotropy

Fluorophores are more likely to be excited by photons if the transition moment of the fluorophore is parallel to the electric vector of the photon. The polarization of the emitted light will also depend on the transition moment. The transition moment is dependent on the physical orientation of the fluorophore molecule. For fluorophores in solution this means that the intensity and polarization of the emitted light is dependent on rotational diffusion. Therefore, anisotropy measurements can be used to investigate how freely a fluorescent molecule moves in a particular environment.

Fluorescence anisotropy can be defined quantitatively as:

$$r = \frac{I_\parallel - I_\perp}{I_\parallel + 2I_\perp}$$

where $I_\parallel$ is the emitted intensity parallel to polarization of the excitation light and $I_\perp$ is the emitted intensity perpendicular to the polarization of the excitation light.

Fluorescence

Strongly fluorescent pigments often have an unusual appearance which is often described colloquially as a "neon color". This phenomenon was termed "Farbenglut" by Hermann von Helmholtz and "fluorence" by Ralph M. Evans. It is generally thought to be related to the high brightness of the color relative to what it would be as a component of white. Fluorescence shifts energy in the incident illumination from shorter wavelengths to longer (such as blue to yellow) and thus can make the fluorescent color appear brighter (more saturated) than it could possibly be by reflection alone.

Rules

There are several general rules that deal with fluorescence. Each of the following rules has exceptions but they are useful guidelines for understanding fluorescence (these rules do not necessarily apply to two-photon absorption).

Kasha's Rule

Kasha's rule dictates that the quantum yield of luminescence is independent of the wavelength of exciting radiation. This occurs because excited molecules usually decay to the lowest vibrational level of the excited state before fluorescence emission takes place. The Kasha–Vavilov rule does not always apply and is violated severely in many simple molecules. A somewhat more reliable statement, although still with exceptions, would be that the fluorescence spectrum shows very little dependence on the wavelength of exciting radiation.

Mirror Image Rule

For many fluorophores the absorption spectrum is a mirror image of the emission spectrum. This is known as the mirror image rule and is related to the Franck–Condon principle which states that electronic transitions are vertical, that is energy changes without distance changing as can be represented with a vertical line in Jablonski diagram. This means the nucleus does not move and the vibration levels of the excited state resemble the vibration levels of the ground state.

Stokes Shift

In general, emitted fluorescence light has a longer wavelength and lower energy than the absorbed light. This phenomenon, known as Stokes shift, is due to energy loss between the time a photon is absorbed and when a new one is emitted. The causes and magnitude of Stokes shift can be complex and are dependent on the fluorophore and its environment. However, there are some common causes. It is frequently due to non-radiative decay to the lowest vibrational energy level of the excited state. Another factor is that the emission of fluorescence frequently leaves a fluorophore in a higher vibrational level of the ground state.

Fluorescence in Nature

There are many natural compounds that exhibit fluorescence, and they have a number of applications. Some deep-sea animals, such as the greeneye, have fluorescent structures.

Fluorescence vs. Bioluminescence vs. Biophosphorescence

Fluorescence

Fluorescence is the temporary absorption of electromagnetic wavelengths from the visible light spectrum by fluorescent molecules, and the subsequent emission of light at a lower energy level. When it occurs in a living organism, it is sometimes called biofluorescence. This causes the light that is emitted to be a different color than the light that is absorbed. Stimulating light excites an electron, raising energy to an unstable level. This instability is unfavorable, so the energized electron is returned to a stable state almost as immediately as it becomes unstable. This return to stability corresponds with the release of excess energy in the form of fluorescence light. This emission of light is only observable when the stimulant light is still providing light to the organism/object and is typically yellow, pink, orange, red, green, or purple. Fluorescence is often confused with the following forms of biotic light, bioluminescence and biophosphorescence. Pumpkin toadlets that live in the Brazilian Atlantic forest are fluorescent.

Bioluminescence

Bioluminescence differs from fluorescence in that it is the natural production of light by chemical reactions within an organism, whereas fluorescence is the absorption and reemission of light from the environment. A firefly and anglerfish are two example of bioluminescent organisms.

Phosphorescence

Biophosphorescence is similar to fluorescence in its requirement of light wavelengths as a provider of excitation energy. The difference here lies in the relative stability of the energized electron. Unlike with fluorescence, in phosphorescence the electron retains stability, emitting light that continues to "glow-in-the-dark" even after the stimulating light source has been removed. Glow-in-the-dark stickers are phosphorescent, but there are no truly phosphorescent animals known.

Mechanisms of Fluorescence

Epidermal Chromatophores

Pigment cells that exhibit fluorescence are called fluorescent chromatophores, and function somatically similar to regular chromatophores. These cells are dendritic, and contain pigments called fluorosomes. These pigments contain fluorescent proteins which are activated by K+ (potassium) ions, and it is their movement, aggregation, and dispersion within the fluorescent chromatophore that cause directed fluorescence patterning. Fluorescent cells are innervated the same as other chromatphores, like melanophores, pigment cells that contain melanin. Short term fluorescent patterning and signaling is controlled by the nervous system. Fluorescent chromatophores can be found in the skin (e.g. in fish) just below the epidermis, amongst other chromatophores.

Epidermal fluorescent cells in fish also respond to hormonal stimuli by the α–MSH and MCH hormones much the same as melanophores. This suggests that fluorescent cells may have color changes throughout the day that coincide with their circadian rhythm. Fish may also be sensitive to cortisol induced stress responses to environmental stimuli, such as interaction with a predator or engaging in a mating ritual.

Phylogenetics

Evolutionary Origins

It is suspected by some scientists that GFPs and GFP like proteins began as electron donors activated by light. These electrons were then used for reactions requiring light energy. Functions of fluorescent proteins, such as protection from the sun, conversion of light into different wavelengths, or for signaling are thought to have evolved secondarily. Fluorescence has multiple origins in the tree of life.

The incidence of fluorescence across the tree of life is widespread, and has been studied most extensively in cnidarians and fish. The phenomenon appears to have evolved multiple times in multiple taxa such as in the anguilliformes (eels), gobioidei (gobies and cardinalfishes), and tetradontiformes (triggerfishes), along with the other taxa discussed later in the article. Fluorescence is highly genotypically and phenotypically variable even within ecosystems, in regards to the wavelengths emitted, the patterns displayed, and the intensity of the fluorescence. Generally, the species relying upon camouflage exhibit the greatest diversity in fluorescence, likely because camouflage may be one of the uses of fluorescence.

Adaptive Functions

Currently, relatively little is known about the functional significance of fluorescence and fluorescent proteins. However, it is suspected that fluorescence may serve important functions in signaling and communication, mating, lures, camouflage, UV protection and antioxidation, photoacclimation, dinoflagellate regulation, and in coral health.

Aquatic Fluorescence

Water absorbs light of long wavelengths, so less light from these wavelengths reflects back to reach the eye. Therefore, warm colors from the visual light spectrum appear less vibrant at increasing depths. Water scatters light of shorter wavelengths above violet, meaning cooler colors dominate the visual field in the photic zone. Light intensity decreases 10 fold with every 75 m of depth, so at depths of 75 m, light is 10% as intense as it is on the surface, and is only 1% as intense at 150 m as it is on the surface. Because the water filters out the wavelengths and intensity of water reaching certain depths, different proteins, because of the wavelengths and intensities of light they are capable of absorbing, are better suited to different depths. Theoretically, some fish eyes can detect light as deep as 1000 m. At these depths of the aphotic zone, the only sources of light are organisms themselves, giving off light through chemical reactions in a process called bioluminescence.

Fluorescence is simply defined as the absorption of electromagnetic radiation at one wavelength and its reemission at another, lower energy wavelength. Thus any type of fluorescence depends on the presence of external sources of light. Biologically functional fluorescence is found in the photic zone, where there is not only enough light to cause fluorescence, but enough light for other organisms to detect it. The visual field in the photic zone is naturally blue, so colors of fluorescence can be detected as bright reds, oranges, yellows, and greens. Green is the most commonly found color in the marine spectrum, yellow the second most, orange the third, and red is the rarest. Fluorescence can occur in organisms in the aphotic zone as a byproduct of that same organism's bioluminescence. Some fluorescence in the aphotic zone is merely a byproduct of the organism's tissue biochemistry and does not have a functional purpose. However, some cases of functional and adaptive significance of fluorescence in the aphotic zone of the deep ocean is an active area of research.

Applications of Fluorescence

Lighting

Fluorescent paint and plastic lit by UV tubes. Paintings by Beo Beyond.

The common fluorescent lamp relies on fluorescence. Inside the glass tube is a partial vacuum and a small amount of mercury. An electric discharge in the tube causes the mercury atoms to emit mostly ultraviolet light. The tube is lined with a coating of a fluorescent material, called the *phosphor*, which absorbs ultraviolet light and re-emits visible light. Fluorescent lighting is more energy-efficient than incandescent lighting elements. However, the uneven spectrum of traditional fluorescent lamps may cause certain colors to appear different than when illuminated by incandescent light or daylight. The mercury vapor emission spectrum is dominated by a short-wave UV line at 254 nm (which provides most of the energy to the phosphors), accompanied by visible light emission at 436 nm (blue), 546 nm (green) and 579 nm (yellow-orange). These three lines can be observed superimposed on the white continuum using a hand spectroscope, for light emitted by the usual white fluorescent tubes. These same visible lines, accompanied by the emission lines of trivalent europium and trivalent terbium, and further accompanied by the emission continuum of divalent europium in the blue region, comprise the more discontinuous light emission of the modern trichromatic phosphor systems used in many compact fluorescent lamp and traditional lamps where better color rendition is a goal.

Fluorescent lights were first available to the public at the 1939 New York World's Fair. Improvements since then have largely been better phosphors, longer life, and more consistent internal discharge, and easier-to-use shapes (such as compact fluorescent lamps). Some high-intensity discharge (HID) lamps couple their even-greater electrical efficiency with phosphor enhancement for better color rendition.

White light-emitting diodes (LEDs) became available in the mid-1990s as LED lamps, in which blue light emitted from the semiconductor strikes phosphors deposited on the tiny chip. The combination of the blue light that continues through the phosphor and the green to red fluorescence from the phosphors produces a net emission of white light.

Glow sticks sometimes utilize fluorescent materials to absorb light from the chemiluminescent reaction and emit light of a different color.

Analytical Chemistry

Many analytical procedures involve the use of a fluorometer, usually with a single exciting wavelength and single detection wavelength. Because of the sensitivity that the method affords, fluorescent molecule concentrations as low as 1 part per trillion can be measured.

Fluorescence in several wavelengths can be detected by an array detector, to detect compounds from HPLC flow. Also, TLC plates can be visualized if the compounds or a coloring reagent is fluorescent. Fluorescence is most effective when there is a larger ratio of atoms at lower energy levels in a Boltzmann distribution. There is, then, a higher probability of excitement and release of photons by lower-energy atoms, making analysis more efficient.

Spectroscopy

Usually the setup of a fluorescence assay involves a light source, which may emit many different wavelengths of light. In general, a single wavelength is required for proper analysis, so, in order to selectively filter the light, it is passed through an excitation monochromator, and then that chosen wavelength is passed through the sample cell. After absorption and re-emission of the energy, many wavelengths may emerge due to Stokes shift and various electron transitions. To separate and analyze them, the fluorescent radiation is passed through an emission monochromator, and observed selectively by a detector.

Biochemistry and Medicine

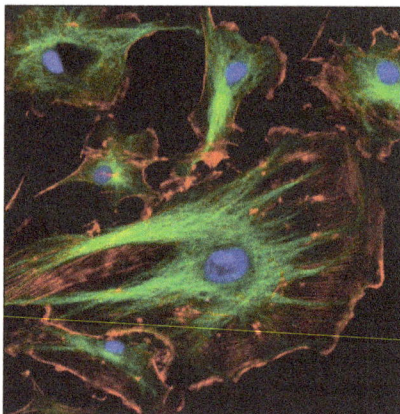

Endothelial cells under the microscope with three separate
channels marking specific cellular components.

Fluorescence in the life sciences is used generally as a non-destructive way of tracking or analysis of biological molecules by means of the fluorescent emission at a specific frequency where there is no background from the excitation light, as relatively few cellular components are naturally fluorescent (called intrinsic or autofluorescence). In fact, a protein or other component can be "labelled" with an extrinsic fluorophore, a fluorescent dye that can be a small molecule, protein, or quantum dot, finding a large use in many biological applications.

The quantification of a dye is done with a spectrofluorometer and finds additional applications in:

Microscopy

- When scanning the fluorescence intensity across a plane one has fluorescence microscopy of tissues, cells, or subcellular structures, which is accomplished by labeling an antibody with a fluorophore and allowing the antibody to find its target antigen within the sample. Labelling multiple antibodies with different fluorophores allows visualization of multiple targets within a single image (multiple channels). DNA microarrays are a variant of this.

- Immunology: An antibody is first prepared by having a fluorescent chemical group attached, and the sites (e.g., on a microscopic specimen) where the antibody has bound can be seen, and even quantified, by the fluorescence.

- FLIM (Fluorescence Lifetime Imaging Microscopy) can be used to detect certain bio-molecular interactions that manifest themselves by influencing fluorescence life-times.

- Cell and molecular biology: detection of colocalization using fluorescence-labelled antibodies for selective detection of the antigens of interest using specialized software such as ImageJ.

Other Techniques

- FRET (Förster resonance energy transfer, also known as fluorescence resonance energy transfer) is used to study protein interactions, detect specific nucleic acid sequences and used as biosensors, while fluorescence lifetime (FLIM) can give an additional layer of information.

- Biotechnology: Biosensors using fluorescence are being studied as possible Fluorescent glucose biosensors.

- Automated sequencing of DNA by the chain termination method; each of four different chain terminating bases has its own specific fluorescent tag. As the labelled DNA molecules are separated, the fluorescent label is excited by a UV source, and the identity of the base terminating the molecule is identified by the wavelength of the emitted light.

- FACS (fluorescence-activated cell sorting): One of several important cell sorting techniques used in the separation of different cell lines (especially those isolated from animal tissues).

- DNA detection: The compound ethidium bromide, in aqueous solution, has very little fluorescence, as it is quenched by water. Ethidium bromide's fluorescence is greatly enhanced after it binds to DNA, so this compound is very useful in visualising the location of DNA fragments in agarose gel electrophoresis. Intercalated ethidium is in a hydrophobic environment when it is between the base pairs of the DNA, protected from quenching by water which is excluded from the local environment of the intercalated ethidium. Ethidium bromide may be carcinogenic – an arguably safer alternative is the dye SYBR Green.

- FIGS (Fluorescence image-guided surgery) is a medical imaging technique that uses fluorescence to detect properly labeled structures during surgery.

- Intravascular fluorescence is a catheter-based medical imaging technique that uses fluorescence to detect high-risk features of atherosclerosis and unhealed vascular stent devices. Plaque autofluorescence has been used in a first-in-man study in coronary arteries in combination with optical coherence tomography. Molecular agents has been also used to detect specific features, such as stent fibrin accumulation and enzymatic activity related to artery inflammation.

- SAFI (species altered fluorescence imaging) an imaging technique in electrokinetics and microfluidics. It uses non-electromigrating dyes whose fluorescence is easily quenched by migrating chemical species of interest. The dye(s) are usually seeded everywhere in the flow and differential quenching of their fluorescence by analytes is directly observed.

- Fluorescence-based assays for screening toxic chemicals. The optical assays consist of a mixture of environmental-sensitive fluorescent dyes and human skin cells that generate fluorescence spectra patterns. This approach can reduce the need for laboratory animals in biomedical research and pharmaceutical industry.

- Bone-margin detection: Alizarin-stained specimens and certain fossils can be lit by fluorescent lights to view anatomical structures, including bone margins.

Forensics

Fingerprints can be visualized with fluorescent compounds such as ninhydrin or DFO (1,8-Diazafluoren-9-one). Blood and other substances are sometimes detected by fluorescent reagents, like fluorescein. Fibers, and other materials that may be encountered in forensics or with a relationship to various collectibles, are sometimes fluorescent.

Non-destructive Testing

Fluorescent penetrant inspection is used to find cracks and other defects on the surface of a part. Dye tracing, using fluorescent dyes, is used to find leaks in liquid and gas plumbing systems.

Signage

Fluorescent colors are frequently used in signage, particularly road signs. Fluorescent colors are generally recognizable at longer ranges than their non-fluorescent counterparts, with fluorescent orange being particularly noticeable. This property has led to its frequent use in safety signs and labels.

Optical Brighteners

Fluorescent compounds are often used to enhance the appearance of fabric and paper, causing a "whitening" effect. A white surface treated with an optical brightener can emit more visible light than that which shines on it, making it appear brighter. The blue light emitted by the brightener compensates for the diminishing blue of the treated material and changes the hue away from yellow or brown and toward white. Optical brighteners are used in laundry detergents, high brightness paper, cosmetics, high-visibility clothing and more.

Photodegradation

Photodegraded plastic bag adjacent to hiking trail.
Appx 2,000 pieces 1 to 25 mm. 3 months exposure outdoors.

Photodegradation is the alteration of materials by light. Typically, the term refers to the combined action of sunlight and air. Photodegradation is usually oxidation and hydrolysis. Often photodegradation is avoided, since it destroys paintings and other artifacts. It is however partly responsible for remineralization of biomass and is used intentionally in some disinfection technologies. Photodegradation does not apply to how materials may be aged or degraded via infrared light or heat, but does include degradation in all of the ultraviolet light wavebands

Applications

Foodstuffs

The protection of food from photodegradation is very important. Some nutrients, for example, are affected by degradation when exposed to sunlight. In the case of beer, UV radiation causes a process that entails the degradation of hop bitter compounds to 3-methyl-2-buten-1-thiol and therefore changes the taste. As amber-colored glass has the ability to absorb UV radiation, beer bottles are often made from such glass to prevent this process.

Paints, Inks and Dyes

Paints, inks and dyes that are organic are more susceptible to photodegradation than those that are not. Ceramics are almost universally colored with non-organic origin materials so as to allow the material to resist photodegradation even under the most relentless conditions, maintaining its color.

Pesticides and Herbicides

The photodegradation of pesticides is of great interest because of the scale of agriculture and the intensive use of chemicals. Pesticides are however selected in part not to photodegrade readily in sunlight in order to allow them to exert their biocidal activity. Thus, additional modalities are implemented to enhance their photodegradation, including the use of photosensitizers, photocatalysts (e.g., titanium dioxide), and the addition of reagents such as hydrogen peroxide that would generate hydroxyl radicals that would attack the pesticides.

Pharmaceuticals

The photodegradation of pharmaceuticals is of interest because they are found in many water supplies. They have deleterious effects on aquatic organisms including toxicity, endocrine disruption, genetic damage. But also in the primary packaging material the photodegradation of pharmaceuticals has to be prevented. For this, amber glasses like Fiolax amber and Corning 51-L are commonly used to protect the pharmaceutical from UV radiations. Iodine (in the form of Lugol's solution) and colloidal silver are universally used in packaging that lets through very little UV light so as to avoid degradation.

Polymers

Common synthetic polymers that can be attacked include polypropylene and LDPE, where tertiary carbon bonds in their chain structures are the centres of attack. Ultraviolet rays interact with these bonds to form free radicals, which then react further with oxygen in the atmosphere, producing carbonyl groups in the main chain. The exposed surfaces of products may then discolour and crack, and in extreme cases, complete product disintegration can occur.

Effect of UV exposure on polypropylene rope.

In fibre products like rope used in outdoor applications, product life will be low because the outer fibres will be attacked first, and will easily be damaged by abrasion for example. Discolouration of the rope may also occur, thus giving an early warning of the problem.

Polymers which possess UV-absorbing groups such as aromatic rings may also be sensitive to UV degradation. Aramid fibres like Kevlar, for example, are highly UV-sensitive and must be protected from the deleterious effects of sunlight.

Mechanism

Photodegradation of a plastic bucket used as an open-air flowerpot for some years.

Many organic chemicals are thermodynamically unstable in the presence of oxygen, however, their rate of spontaneous oxidation is slow at room temperature. In the language of physical chemistry, such reactions are kinetically limited. This kinetic stability allows the accumulation of complex environmental structures in the environment. Upon the absorption of light, triplet oxygen converts to singlet oxygen, a highly reactive form of the gas, which effects spin-allowed oxidations. In the atmosphere, the organic compounds are degraded by hydroxyl radicals, which are produced from water and ozone.

Photochemical reactions are initiated by the absorption of a photon, typically in the wavelength range 290-700 nm (at the surface of the Earth). The energy of an absorbed photon is transferred to electrons in the molecule and briefly changes their configuration (i.e., promotes the molecule from a ground state to an excited state). The excited state represents what is essentially a new molecule. Often excited state molecules are not kinetically stable in the presence of O_2 or H_2O and can spontaneously decompose (oxidize or hydrolyze). Sometimes molecules decompose to produce high energy, unstable fragments that can react with other molecules around them. The two processes are collectively referred to as direct photolysis or indirect photolysis, and both mechanisms contribute to the removal of pollutants.

The United States federal standard for testing plastic for photo-degradation is 40 CFR Ch. I (7–1–03 Edition)PART 238.

Protection against Photodegradation

Photodegradation of plastics and other materials can be inhibited with additives, which are widely used. These additives include antioxidants, which interrupt degradation processes. Typical antioxidants are derivatives of aniline. Another type of additive are UV-absorbers. These agents capture the photon and convert it to heat. Typical UV-absorbers are hydroxy-substituted benzophenones, related to the chemicals used in sunscreen.

Photosensitization

Photosensitization is the process of initiating a reaction through the use of a substance capable of absorbing light and transferring the energy to the desired reactants. The technique is commonly employed in photochemical work, particularly for reactions requiring light sources of certain wavelengths that are not readily available. A commonly used sensitizer is mercury, which absorbs radiation at 1849 and 2537 angstroms; these are the wavelengths of light produced in high-intensity mercury lamps. Also used as sensitizers are cadmium; some of the noble gases, particularly xenon; zinc; benzophenone; and a large number of organic dyes.

In a typical photosensitized reaction, as in the photodecomposition of ethylene to acetylene and hydrogen, a mixture of mercury vapour and ethylene is irradiated with a mercury lamp. The mercury atoms absorb the light energy, there being a suitable electronic transition in the atom that corresponds to the energy of the incident light. In colliding with ethylene molecules, the mercury atoms transfer the energy and are in turn deactivated to their initial energy state. The excited ethylene molecules subsequently undergo decomposition. Another mode of photosensitization observed in many reactions involves direct participation of the sensitizer in the reaction itself.

Photodissociation

Photodissociation, photolysis, or photodecomposition is a chemical reaction in which a chemical compound is broken down by photons. It is defined as the interaction of one or more photons with one target molecule. Photodissociation is not limited to visible light. Any photon with sufficient energy can affect the chemical bonds of a chemical compound. Since a photon's energy is inversely proportional to its wavelength, electromagnetic waves with the energy of visible light or higher, such as ultraviolet light, x-rays and gamma rays are usually involved in such reactions.

Photolysis in Photosynthesis

Photolysis is part of the light-dependent reactions of photosynthesis. The general reaction of photosynthetic photolysis can be given as

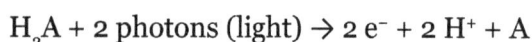

$$H_2A + 2 \text{ photons (light)} \rightarrow 2\,e^- + 2\,H^+ + A$$

The chemical nature of "A" depends on the type of organism. In purple sulfur bacteria, hydrogen sulfide (H_2S) is oxidized to sulfur (S). In oxygenic photosynthesis, water (H_2O) serves as a substrate for photolysis resulting in the generation of diatomic oxygen (O_2). This is the process which returns oxygen to Earth's atmosphere. Photolysis of water occurs in the thylakoids of cyanobacteria and the chloroplasts of green algae and plants.

Energy Transfer Models

The conventional, semi-classical, model describes the photosynthetic energy transfer process as one in which excitation energy hops from light-capturing pigment molecules to reaction center molecules step-by-step down the molecular energy ladder.

The effectiveness of photons of different wavelengths depends on the absorption spectra of the photosynthetic pigments in the organism. Chlorophylls absorb light in the violet-blue and red parts of the spectrum, while accessory pigments capture other wavelengths as well. The phycobilins of red algae absorb blue-green light which penetrates deeper into water than red light, enabling them to photosynthesize in deep waters. Each absorbed photon causes the formation of an exciton (an electron excited to a higher energy state) in the pigment molecule. The energy of the exciton is transferred to a chlorophyll molecule (P680, where P stands for pigment and 680 for its absorption maximum at 680 nm) in the reaction center of photosystem II via resonance energy transfer. P680 can also directly absorb a photon at a suitable wavelength.

Photolysis during photosynthesis occurs in a series of light-driven oxidation events. The energized electron (exciton) of P680 is captured by a primary electron acceptor of the photosynthetic electron transfer chain and thus exits photosystem II. In order to repeat the reaction, the electron in the reaction center needs to be replenished. This occurs by oxidation of water in the case of oxygenic photosynthesis. The electron-deficient reaction center of photosystem II (P680*) is the strongest biological oxidizing agent yet discovered, which allows it to break apart molecules as stable as water.

The water-splitting reaction is catalyzed by the oxygen evolving complex of photosystem II. This protein-bound inorganic complex contains four manganese ions, plus calcium and chloride ions as cofactors. Two water molecules are complexed by the manganese cluster, which then undergoes a series of four electron removals (oxidations) to replenish the reaction center of photosystem II. At the end of this cycle, free oxygen (O_2) is generated and the hydrogen of the water molecules has been converted to four protons released into the thylakoid lumen.

These protons, as well as additional protons pumped across the thylakoid membrane coupled with the electron transfer chain, form a proton gradient across the membrane that drives photophosphorylation and thus the generation of chemical energy in the form of adenosine tri-phosphate (ATP). The electrons reach the P700 reaction center of photosystem I where they are energized again by light. They are passed down another electron transfer chain and finally combine with the coenzyme NADP$^+$ and protons outside the thylakoids to form NADPH. Thus, the net oxidation reaction of water photolysis can be written as:

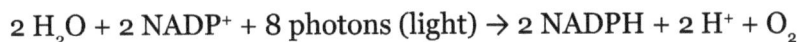

$$2\ H_2O + 2\ NADP^+ + 8\ photons\ (light) \rightarrow 2\ NADPH + 2\ H^+ + O_2$$

The free energy change (ΔG) for this reaction is 102 kilocalories per mole. Since the energy of light at 700 nm is about 40 kilocalories per mole of photons, approximately 320 kilocalories of light energy are available for the reaction. Therefore, approximately one-third of the available light energy is captured as NADPH during photolysis and electron transfer. An equal amount of ATP is generated by the resulting proton gradient. Oxygen as a byproduct is of no further use to the reaction and thus released into the atmosphere.

Quantum Models

In 2007 a quantum model was proposed by Graham Fleming and his co-workers which includes the possibility that photosynthetic energy transfer might involve quantum oscillations, explaining its unusually high efficiency.

According to Fleming there is direct evidence that remarkably long-lived wavelike electronic quantum coherence plays an important part in energy transfer processes during photosynthesis, which can explain the extreme efficiency of the energy transfer because it enables the system to sample all the potential energy pathways, with low loss, and choose the most efficient one.

This approach has been further investigated by Gregory Scholes and his team at the University of Toronto, which in early 2010 published research results that indicate that some marine algae make use of quantum-coherent electronic energy transfer (EET) to enhance the efficiency of their energy harnessing.

Photoinduced Proton Transfer

Photoacids are molecules that upon light absorption undergo a proton transfer to form the photobase.

$$AH \xrightarrow{\ hn\ } A^- + H^+$$

In these reactions the dissociation occurs in the electronically excited state. After proton transfer and relaxation to the electronic ground state, the proton and acid recombine to form the photoacid again.

Photoacids are a convenient source to induce pH jumps in ultrafast laser spectroscopy experiments.

Photolysis in the Atmosphere

Photolysis occurs in the atmosphere as part of a series of reactions by which primary pollutants such as hydrocarbons and nitrogen oxides react to form secondary pollutants such as peroxyacyl nitrates.

The two most important photodissociaton reactions in the troposphere are firstly:

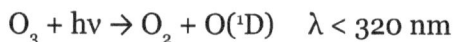

$$O_3 + hv \rightarrow O_2 + O(^1D) \quad \lambda < 320 \text{ nm}$$

which generates an excited oxygen atom which can react with water to give the hydroxyl radical:

$$O(^1D) + H_2O \rightarrow 2 \ ^.OH$$

The hydroxyl radical is central to atmospheric chemistry as it initiates the oxidation of hydrocarbons in the atmosphere and so acts as a detergent.

Secondly the reaction:

$$NO_2 + hv \rightarrow NO + O$$

is a key reaction in the formation of tropospheric ozone.

The formation of the ozone layer is also caused by photodissociation. Ozone in the Earth's stratosphere is created by ultraviolet light striking oxygen molecules containing two oxygen atoms (O_2), splitting them into individual oxygen atoms (atomic oxygen). The atomic oxygen then combines with unbroken O_2 to create ozone, O_3. In addition, photolysis is the process by which CFCs are broken down in the upper atmosphere to form ozone-destroying chlorine free radicals.

Astrophysics

In astrophysics, photodissociation is one of the major processes through which molecules are broken down (but new molecules are being formed). Because of the vacuum of the interstellar medium, molecules and free radicals can exist for a long time. Photodissociation is the main path by which molecules are broken down. Photodissociation rates are important in the study of the composition of interstellar clouds in which stars are formed.

Examples of photodissociation in the interstellar medium are (*hv* is the energy of a single photon of frequency *v*):

$$H_2O \xrightarrow{hv} H + OH$$

$$CH_4 \xrightarrow{hv} CH_3 + H$$

Atmospheric Gamma-ray Bursts

Currently orbiting satellites detect an average of about one gamma-ray burst per day. Because gamma-ray bursts are visible to distances encompassing most of the observable universe, a volume encompassing many billions of galaxies, this suggests that gamma-ray bursts must be exceedingly rare events per galaxy.

Measuring the exact rate of gamma-ray bursts is difficult, but for a galaxy of approximately the same size as the Milky Way, the expected rate (for long GRBs) is about one burst every 100,000 to 1,000,000 years. Only a few percent of these would be beamed towards Earth. Estimates of rates of short GRBs are even more uncertain because of the unknown beaming fraction, but are probably comparable.

A gamma-ray burst in the Milky Way, if close enough to Earth and beamed towards it, could have significant effects on the biosphere. The absorption of radiation in the atmosphere would cause photodissociation of nitrogen, generating nitric oxide that would act as a catalyst to destroy ozone.

The atmospheric photodissociation:

- $N_2 \rightarrow 2N$

- $O_2 \rightarrow 2O$

- $CO_2 \rightarrow C + 2O$

- $H_2O \rightarrow 2H + O$

- $2NH_3 \rightarrow 3H_2 + N_2$

would yield:

- NO_2 (consumes up to 400 ozone molecules)

- CH_2 (nominal)

- CH_4 (nominal)

- CO_2 (incomplete)

According to a 2004 study, a GRB at a distance of about a kiloparsec could destroy up to half of Earth's ozone layer; the direct UV irradiation from the burst combined with additional solar UV radiation passing through the diminished ozone layer could then have potentially significant impacts on the food chain and potentially trigger a mass extinction. The authors estimate that one such burst is expected per billion years, and hypothesize that the Ordovician-Silurian extinction event could have been the result of such a burst.

There are strong indications that long gamma-ray bursts preferentially or exclusively occur in regions of low metallicity. Because the Milky Way has been metal-rich since before the Earth formed, this effect may diminish or even eliminate the possibility that a long gamma-ray burst has occurred within the Milky Way within the past billion years. No such metallicity biases are known for short gamma-ray bursts. Thus, depending on their local rate and beaming properties, the possibility for a nearby event to have had a large impact on Earth at some point in geological time may still be significant.

Multiple Photon Dissociation

Single photons in the infrared spectral range usually are not energetic enough for direct photodissociation of molecules. However, after absorption of multiple infrared photons a molecule may gain internal energy to overcome its barrier for dissociation. Multiple photon dissociation (MPD, IRMPD with infrared radiation) can be achieved by applying high power lasers, e.g. a carbon dioxide laser, or a free electron laser, or by long interaction times of the molecule with the radiation field without the possibility for rapid cooling, e.g. by collisions. The latter method allows even for MPD induced by black-body radiation, a technique called blackbody infrared radiative dissociation (BIRD).

Photorearrangement

In photorearrangement, absorption of light causes a molecule to rearrange its structure in such a way that atoms are lost and it becomes another chemical species. One biologically important photorearrangement reaction is the conversion of 7-dehydrocholesterol to vitamin D in the skin. Lack of exposure to solar radiation can cause a deficiency of vitamin D, which leads to a debilitating decalcification of the bones called rickets. This disorder was first described by Roman physicians in the 2nd century BCE, and, at the height of the Industrial Revolution, it affected 90 percent of children raised in the crowded cities of Europe and North America. Early in the 19th century it was recognized that rickets could be prevented by exposure to sunlight, and this practice became widely adopted at the beginning of the 20th century as an effective treatment.

Plants in the human diet contribute 7-dehydrocholesterol, which accumulates in cholesterol-rich rafts in the plasma membrane of skin cells. While in the skin, 7-dehydrocholesterol absorbs UV light (about 300 nm), leading to the photorearrangement. In this reaction the bond between one carbon and one hydrogen atom is eliminated, while simultaneously the same hydrogen atom forms a bond to a new carbon atom, resulting in the molecule cholecalciferol, or vitamin D_3.

Though it is not biologically active itself, cholecalciferol is converted by the liver and the kidneys into several forms of vitamin D with various metabolic roles, including regulating calcium (Ca^{2+}) levels in the intestine, kidney, liver, and bone and controlling differentiation of hematopoetic cells in bone marrow to macrophages and osteoclasts for bone formation. It is also an antiproliferative agent for breast and colon carcinomas, lymphomas, and leukemias.

Photoisomerization

In photoisomerization no chemical bonds are broken, but the molecule changes shape. For example, absorption of optical radiation by a stilbene molecule converts the central double bond from *trans* to *cis*. As in photodissociation, this is caused by the electron distribution in the excited state being quite different from that in the ground state; hence, the structure of the initially created excited singlet (by absorption of light) is most stable at 90°, or halfway between the *cis* and *trans* forms. The molecule attempts to adopt this conformation by rotating about the double bond until the shape of its nuclei matches the distribution of its electrons. Internal conversion occurs most efficiently from this point where the S_0 and S_1 energies are close. Thus, within one or a few molecular vibrations (30–100 fs), the molecule returns to the S_0 state with excess vibrational energy. However, the 90° twist of the double bond is the least-stable conformation for the electron distribution of the S_0 state, so the molecule again rotates about the double bond. Rotation can either continue in the same direction, forming the new isomer, or go back, forming the original isomer. In reality the motions of the molecule are more complicated than described here, involving simultaneous rotation about multiple bonds. However, this simple description contains the essence of the process.

The primary step in vision is the photoisomerization of a retinol (vitamin A) molecule bound within a specialized protein (opsin). The visual pigment (e.g., retinal) and the protein together constitute one of a large family of membrane-bound photoreceptors, or rhodopsins. These protein-pigment complexes are responsible for all of the body's responses to light, including vision, growth and division of melanocytes (tanning), regulation of circadian rhythms (the body's 24-hour cycle), opening and closing of the iris, and others. The active centre of rhodopsin is found in rod cells of the retina. The retinal molecule has several conjugated double bonds, which are all *trans* except for one in the *cis* conformation. This single *cis* bond photoisomerizes rapidly and efficiently to *trans*, driving a change in the protein structure that then initiates a cascade of events leading eventually to a nerve impulse.

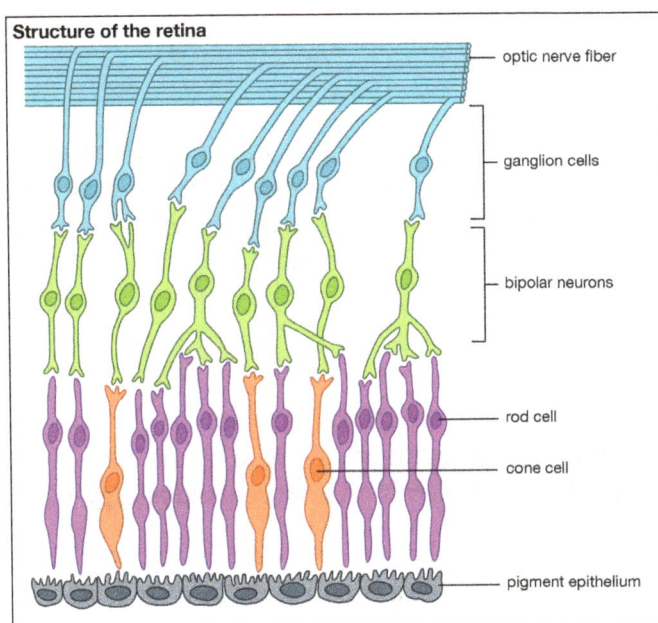

Structure of the retina.

Rod cells are the most sensitive to light, but all absorb at the same wavelength, which does not allow colours to be distinguished. In contrast, there are three types of cone cells, each containing a different rhodopsin that absorbs at a slightly different wavelength, enabling colour vision. Remarkably, all cones and rods contain the same retinal chromophore; small differences in the protein shift the rhodopsin absorption (the energy difference between S_1 and S_0) to different colours. In fact, all known animal photoreceptors use retinal as their chromophore. It absorbs light strongly, and, when incorporated into protein, its absorption matches the solar spectrum closely, so it is sensitive in very low light. Also, it is quite stable, so spontaneous isomerization, which would cause false images, almost never occurs. The structural change in the protein upon isomerization is quite large.

References

- Anctil, michel (2018). Luminous creatures: the history and science of light production in living organisms. Montreal & kingston, london, chicago: mcgill-queen's university press. Isbn 978-0-7735-5312-5

- Photochemical-reaction, chemistry: byjus.com, Retrieved 30 April, 2019

- Reshetiloff, kathy (1 july 2001). "chesapeake bay night-lights add sparkle to woods, water". Bay journal. Retrieved 16 december 2014

- Luminescence: scienceclarified.com, Retrieved 29 June, 2019

- Shah, syed niaz ali; lin, jin-ming (2017). "recent advances in chemiluminescence based on carbonaceous dots". Advances in colloid and interface science. 241: 24–36. Doi:10.1016/j.cis.2017.01.003. Pmid 28139217

3
Photosynthesis

The process used by plants and other organisms in which light energy is used to produce glucose from carbon dioxide and water is referred to as photosynthesis. This glucose is then converted into adenosine triphosphate to provide energy. This chapter has been carefully written to provide an easy understanding of photosynthesis.

Photosynthesis is the process by which green plants and certain other organisms transform light energy into chemical energy. During photosynthesis in green plants, light energy is captured and used to convert water, carbon dioxide, and minerals into oxygen and energy-rich organic compounds.

It would be impossible to overestimate the importance of photosynthesis in the maintenance of life on Earth. If photosynthesis ceased, there would soon be little food or other organic matter on Earth. Most organisms would disappear, and in time Earth's atmosphere would become nearly devoid of gaseous oxygen. The only organisms able to exist under such conditions would be the chemosynthetic bacteria, which can utilize the chemical energy of certain inorganic compounds and thus are not dependent on the conversion of light energy.

Energy produced by photosynthesis carried out by plants millions of years ago is responsible for the fossil fuels (i.e., coal, oil, and gas) that power industrial society. In past ages, green plants and small organisms that fed on plants increased faster than they were consumed, and their remains were deposited in Earth's crust by sedimentation and other geological processes. There, protected from oxidation, these organic remains were slowly converted to fossil fuels. These fuels not only provide much of the energy used in factories, homes, and transportation but also serve as the raw material for plastics and other synthetic products. Unfortunately, modern civilization is using up in a few centuries the excess of photosynthetic production accumulated over millions of years. Consequently, the carbon dioxide that has been removed from the air to make carbohydrates in photosynthesis over millions of years is being returned at an incredibly rapid rate. The carbon dioxide concentration in Earth's atmosphere is rising the fastest it ever has in Earth's history, and this phenomenon is expected to have major implications on Earth's climate.

Requirements for food, materials, and energy in a world where human population is rapidly growing have created a need to increase both the amount of photosynthesis and the efficiency of converting photosynthetic output into products useful to people. One response to those needs—the so-called Green Revolution, begun in the mid-20th century—achieved enormous improvements in agricultural yield through the use of chemical fertilizers, pest and plant-disease control, plant breeding, and mechanized tilling, harvesting, and crop processing. This effort limited severe famines to a few areas of the world despite rapid population growth, but it did not eliminate widespread malnutrition. Moreover, beginning in the early 1990s, the rate at which yields of major crops increased began to decline. This was especially true for rice in Asia. Rising costs associated with sustaining high rates of agricultural production, which required ever-increasing inputs of fertilizers and pesticides and constant development of new plant varieties, also became problematic for farmers in many countries.

A second agricultural revolution, based on plant genetic engineering, was forecast to lead to increases in plant productivity and thereby partially alleviate malnutrition. Since the 1970s, molecular biologists have possessed the means to alter a plant's genetic material (deoxyribonucleic acid, or DNA) with the aim of achieving improvements in disease and drought resistance, product yield and quality, frost hardiness, and other desirable properties. However, such traits are inherently complex, and the process of making changes to crop plants through genetic engineering has turned out to be more complicated than anticipated. In the future such genetic engineering may result in improvements in the process of photosynthesis, but by the first decades of the 21st century, it had yet to demonstrate that it could dramatically increase crop yields.

Another intriguing area in the study of photosynthesis has been the discovery that certain animals are able to convert light energy into chemical energy. The emerald green sea slug (*Elysia chlorotica*), for example, acquires genes and chloroplasts from *Vaucheria litorea*, an alga it consumes, giving it a limited ability to produce chlorophyll. When enough chloroplasts are assimilated, the slug may forgo the ingestion of food. The pea aphid (*Acyrthosiphon pisum*) can harness light to manufacture the energy-rich compound adenosine triphosphate (ATP); this ability has been linked to the aphid's manufacture of carotenoid pigments.

General Characteristics

Overall Reaction of Photosynthesis

In chemical terms, photosynthesis is a light-energized oxidation–reduction process. (Oxidation refers to the removal of electrons from a molecule; reduction refers to the gain of electrons by a molecule.) In plant photosynthesis, the energy of light is used to drive the oxidation of water (H_2O), producing oxygen gas (O_2), hydrogen ions (H^+), and electrons. Most of the removed electrons and hydrogen ions ultimately are transferred to carbon dioxide (CO_2), which is reduced to organic products. Other electrons and hydrogen ions are used to reduce nitrate and sulfate to amino and sulfhydryl groups in amino acids, which are the building blocks of proteins. In most green cells, carbohydrates—especially starch and the sugar sucrose—are the major direct organic products of photosynthesis. The overall reaction in which carbohydrates—represented by the general formula (CH_2O)—are formed during plant photosynthesis can be indicated by the following equation:

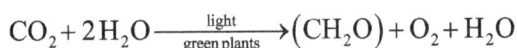

$$CO_2 + 2\,H_2O \xrightarrow[\text{green plants}]{\text{light}} (CH_2O) + O_2 + H_2O$$

This equation is merely a summary statement, for the process of photosynthesis actually involves numerous reactions catalyzed by enzymes (organic catalysts). These reactions occur in two stages: the "light" stage, consisting of photochemical (i.e., light-capturing) reactions; and the "dark" stage, comprising chemical reactions controlled by enzymes. During the first stage, the energy of light is absorbed and used to drive a series of electron transfers, resulting in the synthesis of ATP and the electron-donor-reduced nicotine adenine dinucleotide phosphate (NADPH). During the dark stage, the ATP and NADPH formed in the light-capturing reactions are used to reduce carbon dioxide to organic carbon compounds. This assimilation of inorganic carbon into organic compounds is called carbon fixation.

During the 20th century, comparisons between photosynthetic processes in green plants and in certain photosynthetic sulfur bacteria provided important information about the photosynthetic mechanism. Sulfur bacteria use hydrogen sulfide (H_2S) as a source of hydrogen atoms and produce sulfur instead of oxygen during photosynthesis. The overall reaction is:

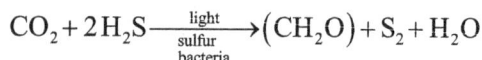

$$CO_2 + 2H_2S \xrightarrow[\substack{sulfur \\ bacteria}]{light} (CH_2O) + S_2 + H_2O$$

In the 1930s Dutch biologist Cornelis van Niel recognized that the utilization of carbon dioxide to form organic compounds was similar in the two types of photosynthetic organisms. Suggesting that differences existed in the light-dependent stage and in the nature of the compounds used as a source of hydrogen atoms, he proposed that hydrogen was transferred from hydrogen sulfide (in bacteria) or water (in green plants) to an unknown acceptor (called A), which was reduced to H_2A. During the dark reactions, which are similar in both bacteria and green plants, the reduced acceptor (H_2A) reacted with carbon dioxide (CO_2) to form carbohydrate (CH_2O) and to oxidize the unknown acceptor to A. This putative reaction can be represented as:

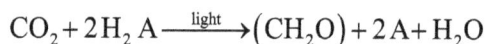

$$CO_2 + 2H_2A \xrightarrow{light} (CH_2O) + 2A + H_2O$$

Van Niel's proposal was important because the popular (but incorrect) theory had been that oxygen was removed from carbon dioxide (rather than hydrogen from water, releasing oxygen) and that carbon then combined with water to form carbohydrate (rather than the hydrogen from water combining with CO_2 to form CH_2O).

By 1940 chemists were using heavy isotopes to follow the reactions of photosynthesis. Water marked with an isotope of oxygen (^{18}O) was used in early experiments. Plants that photosynthesized in the presence of water containing $H_2^{18}O$ produced oxygen gas containing ^{18}O; those that photosynthesized in the presence of normal water produced normal oxygen gas. These results provided definitive support for van Niel's theory that the oxygen gas produced during photosynthesis is derived from water.

Basic Products of Photosynthesis

As has been stated, carbohydrates are the most-important direct organic product of photosynthesis in the majority of green plants. The formation of a simple carbohydrate, glucose, is indicated by a chemical equation,

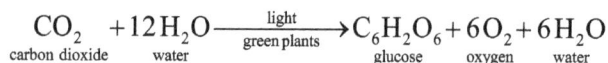

$$\underset{\text{carbon dioxide}}{CO_2} + \underset{\text{water}}{12H_2O} \xrightarrow{\underset{\text{green plants}}{light}} \underset{\text{glucose}}{C_6H_2O_6} + \underset{\text{oxygen}}{6O_2} + \underset{\text{water}}{6H_2O}$$

Little free glucose is produced in plants; instead, glucose units are linked to form starch or are joined with fructose, another sugar, to form sucrose.

Not only carbohydrates, as was once thought, but also amino acids, proteins, lipids (or fats), pigments, and other organic components of green tissues are synthesized during photosynthesis.

Minerals supply the elements (e.g., nitrogen, N; phosphorus, P; sulfur, S) required to form these compounds. Chemical bonds are broken between oxygen (O) and carbon (C), hydrogen (H), nitrogen, and sulfur, and new bonds are formed in products that include gaseous oxygen (O_2) and organic compounds. More energy is required to break the bonds between oxygen and other elements (e.g., in water, nitrate, and sulfate) than is released when new bonds form in the products. This difference in bond energy accounts for a large part of the light energy stored as chemical energy in the organic products formed during photosynthesis. Additional energy is stored in making complex molecules from simple ones.

Evolution of the Process

Although life and the quality of the atmosphere today depend on photosynthesis, it is likely that green plants evolved long after the first living cells. When Earth was young, electrical storms and solar radiation probably provided the energy for the synthesis of complex molecules from abundant simpler ones, such as water, ammonia, and methane. The first living cells probably evolved from these complex molecules. For example, the accidental joining (condensation) of the amino acid glycine and the fatty acid acetate may have formed complex organic molecules known as porphyrins. These molecules, in turn, may have evolved further into coloured molecules called pigments—e.g., chlorophylls of green plants, bacteriochlorophyll of photosynthetic bacteria, hemin (the red pigment of blood), and cytochromes, a group of pigment molecules essential in both photosynthesis and cellular respiration.

Primitive coloured cells then had to evolve mechanisms for using the light energy absorbed by their pigments. At first, the energy may have been used immediately to initiate reactions useful to the cell. As the process for utilization of light energy continued to evolve, however, a larger part of the absorbed light energy probably was stored as chemical energy, to be used to maintain life. Green plants, with their ability to use light energy to convert carbon dioxide and water to carbohydrates and oxygen, are the culmination of this evolutionary process.

The first oxygenic (oxygen-producing) cells probably were the blue-green algae (cyanobacteria), which appeared about two billion to three billion years ago. These microscopic organisms are believed to have greatly increased the oxygen content of the atmosphere, making possible the development of aerobic (oxygen-using) organisms. Cyanophytes are prokaryotic cells; that is, they contain no distinct membrane-enclosed subcellular particles (organelles), such as nuclei and chloroplasts. Green plants, by contrast, are composed of eukaryotic cells, in which the photosynthetic apparatus is contained within membrane-bound chloroplasts. The complete genome sequences of cyanobacteria and higher plants provide evidence that the first photosynthetic eukaryotes were likely the red algae that developed when nonphotosynthetic eukaryotic cells engulfed cyanobacteria. Within the host cells, these cyanobacteria evolved into chloroplasts.

There are a number of photosynthetic bacteria that are not oxygenic (e.g., the sulfur bacteria previously discussed). The evolutionary pathway that led to these bacteria diverged from the one that resulted in oxygenic organisms. In addition to the absence of oxygen production, nonoxygenic photosynthesis differs from oxygenic photosynthesis in two other ways: light of longer wavelengths is absorbed and used by pigments called bacteriochlorophylls, and reduced compounds other than water (such as hydrogen sulfide or organic molecules) provide the electrons needed for the reduction of carbon dioxide.

Factors that Influence the Rate of Photosynthesis

The rate of photosynthesis is defined in terms of the rate of oxygen production either per unit mass (or area) of green plant tissues or per unit weight of total chlorophyll. The amount of light, the carbon dioxide supply, temperature, water supply, and the availability of minerals are the most important environmental factors that affect the rate of photosynthesis in land plants. The rate of photosynthesis is also determined by the plant species and its physiological state—e.g., its health, its maturity, and whether it is in flower.

Light Intensity and Temperature

As has been mentioned, the complex mechanism of photosynthesis includes a photochemical, or light-harvesting, stage and an enzymatic, or carbon-assimilating, stage that involves chemical reactions. These stages can be distinguished by studying the rates of photosynthesis at various degrees of light saturation (i.e., intensity) and at different temperatures. Over a range of moderate temperatures and at low to medium light intensities (relative to the normal range of the plant species), the rate of photosynthesis increases as the intensity increases and is relatively independent of temperature. As the light intensity increases to higher levels, however, the rate becomes saturated; light "saturation" is achieved at a specific light intensity, dependent on species and growing conditions. In the light-dependent range before saturation, therefore, the rate of photosynthesis is determined by the rates of photochemical steps. At high light intensities, some of the chemical reactions of the dark stage become rate-limiting. In many land plants, a process called photorespiration occurs, and its influence upon photosynthesis increases with rising temperatures. More specifically, photorespiration competes with photosynthesis and limits further increases in the rate of photosynthesis, especially if the supply of water is limited.

Carbon Dioxide

Included among the rate-limiting steps of the dark stage of photosynthesis are the chemical reactions by which organic compounds are formed by using carbon dioxide as a carbon source. The rates of these reactions can be increased somewhat by increasing the carbon dioxide concentration. Since the middle of the 19th century, the level of carbon dioxide in the atmosphere has been rising because of the extensive combustion of fossil fuels, cement production, and land-use changes associated with deforestation. The atmospheric level of carbon dioxide climbed from about 0.028 percent in 1860 to 0.032 percent by 1958 (when improved measurements began) and to 0.040 percent by 2016. This increase in carbon dioxide directly increases plant photosynthesis, but the size of the increase depends on the species and physiological condition of the plant. Furthermore, most scientists maintain that increasing levels of atmospheric carbon dioxide affect climate, increasing global temperatures and changing rainfall patterns. Such changes will also affect photosynthesis rates.

Water

For land plants, water availability can function as a limiting factor in photosynthesis and plant growth. Besides the requirement for a small amount of water in the photosynthetic reaction itself, large amounts of water are transpired from the leaves; that is, water evaporates from the leaves to the atmosphere via the stomata. Stomata are small openings through the leaf epidermis, or outer skin; they permit the entry of carbon dioxide but inevitably also allow the exit of water vapour. The stomata open and close according to the physiological needs of the leaf. In hot and arid climates the stomata may close to conserve water, but this closure limits the entry of carbon dioxide and hence the rate of photosynthesis. The decreased transpiration means there is less cooling of the leaves and hence leaf temperatures rise. The decreased carbon dioxide concentration inside the leaves and the increased leaf temperatures favour the wasteful process of photorespiration. If the level of carbon dioxide in the atmosphere increases, more carbon dioxide could enter through a smaller opening of the stomata, so more photosynthesis could occur with a given supply of water.

Minerals

Several minerals are required for healthy plant growth and for maximum rates of photosynthesis. Nitrogen, sulfate, phosphate, iron, magnesium, calcium, and potassium are required in substantial amounts for the synthesis of amino acids, proteins, coenzymes, deoxyribonucleic acid (DNA) and ribonucleic acid (RNA), chlorophyll and other pigments, and other essential plant constituents. Smaller amounts of such elements as manganese, copper, and chloride are required in photosynthesis. Some other trace elements are needed for various nonphotosynthetic functions in plants.

Internal Factors

Each plant species is adapted to a range of environmental factors. Within this normal range of conditions, complex regulatory mechanisms in the plant's cells adjust the activities of enzymes (i.e., organic catalysts). These adjustments maintain a balance in the overall photosynthetic process and control it in accordance with the needs of the whole plant. With a given plant species, for example, doubling the carbon dioxide level might cause a temporary increase of nearly twofold in the rate of photosynthesis; a few hours or days later, however, the rate might fall to the original level because photosynthesis produced more sucrose than the rest of the plant could use. By contrast, another plant species provided with such carbon dioxide enrichment might be able to use more sucrose, because it had more carbon-demanding organs, and would continue to photosynthesize and to grow faster throughout most of its life cycle.

Energy Efficiency of Photosynthesis

The energy efficiency of photosynthesis is the ratio of the energy stored to the energy of light absorbed. The chemical energy stored is the difference between that contained in gaseous oxygen and organic compound products and the energy of water, carbon dioxide, and other reactants. The amount of energy stored can only be estimated because many products are formed, and these vary with the plant species and environmental conditions. If the equation for glucose formation given earlier is used to approximate the actual storage process, the production of one mole (i.e., 6.02×10^{23} molecules; abbreviated N) of oxygen and one-sixth mole of glucose results in the storage of about 117 kilocalories (kcal) of chemical energy. This amount must then be compared with the energy of light absorbed to produce one mole of oxygen in order to calculate the efficiency of photosynthesis.

Light can be described as a wave of particles known as photons; these are units of energy, or light quanta. The quantity N photons is called an einstein. The energy of light varies inversely with the length of the photon waves; that is, the shorter the wavelength, the greater the energy content. The energy (e) of a photon is given by the equation $e = hc/\lambda$, where c is the velocity of light, h is Planck's constant, and λ is the light wavelength. The energy (E) of an einstein is $E = Ne = Nhc/\lambda = 28,600/\lambda$, when E is in kilocalories and λ is given in nanometres (nm; 1 nm = 10^{-9} metres). An einstein of red light with a wavelength of 680 nm has an energy of about 42 kcal. Blue light has a shorter wavelength and therefore more energy than red light. Regardless of whether the light is blue or red, however, the same number of einsteins are required for photosynthesis per mole of oxygen formed. The part of the solar spectrum used by plants has an estimated mean wavelength of 570 nm; therefore, the energy of light used during photosynthesis is approximately 28,600/570, or 50 kcal per einstein.

In order to compute the amount of light energy involved in photosynthesis, one other value is needed: the number of einsteins absorbed per mole of oxygen evolved. This is called the quantum requirement. The minimum quantum requirement for photosynthesis under optimal conditions is about nine. Thus, the energy used is 9 × 50, or 450 kcal per mole of oxygen evolved. Therefore, the estimated maximum energy efficiency of photosynthesis is the energy stored per mole of oxygen evolved, 117 kcal, divided by 450—that is, 117/450, or 26 percent.

The actual percentage of solar energy stored by plants is much less than the maximum energy efficiency of photosynthesis. An agricultural crop in which the biomass (total dry weight) stores as much as 1 percent of total solar energy received on an annual areawide basis is exceptional, although a few cases of higher yields (perhaps as much as 3.5 percent in sugarcane) have been reported. There are several reasons for this difference between the predicted maximum efficiency of photosynthesis and the actual energy stored in biomass. First, more than half of the incident sunlight is composed of wavelengths too long to be absorbed, and some of the remainder is reflected or lost to the leaves. Consequently, plants can at best absorb only about 34 percent of the incident sunlight. Second, plants must carry out a variety of physiological processes in such nonphotosynthetic tissues as roots and stems; these processes, as well as cellular respiration in all parts of the plant, use up stored energy. Third, rates of photosynthesis in bright sunlight sometimes exceed the needs of the plants, resulting in the formation of excess sugars and starch. When this happens, the regulatory mechanisms of the plant slow down the process of photosynthesis, allowing more absorbed sunlight to go unused. Fourth, in many plants, energy is wasted by the process of photorespiration. Finally, the growing season may last only a few months of the year; sunlight received during other seasons is not used. Furthermore, it should be noted that if only agricultural products (e.g., seeds, fruits, and tubers, rather than total biomass) are considered as the end product of the energy-conversion process of photosynthesis, the efficiency falls even further.

Chloroplasts: The Photosynthetic Units of Green Plants

The process of plant photosynthesis takes place entirely within the chloroplasts. Detailed studies of the role of these organelles date from the work of British biochemist Robert Hill. About 1940 Hill discovered that green particles obtained from broken cells could produce oxygen from water in the presence of light and a chemical compound, such as ferric oxalate, able to serve as an electron acceptor. This process is known as the Hill reaction. During the 1950s Daniel Arnon and other American biochemists prepared plant cell fragments in which not only the Hill reaction but also the synthesis of the energy-storage compound ATP occurred. In addition, the coenzyme NADP was used as the final acceptor of electrons, replacing the nonphysiological electron acceptors used by Hill. His procedures were refined further so that small individual pieces of isolated chloroplast membranes, or lamellae, could perform the Hill reaction. These small pieces of lamellae were then fragmented into pieces so small that they performed only the light reactions of the photosynthetic process. It is now possible also to isolate the entire chloroplast so that it can carry out the complete process of photosynthesis, from light absorption, oxygen formation, and the reduction of carbon dioxide to the formation of glucose and other products.

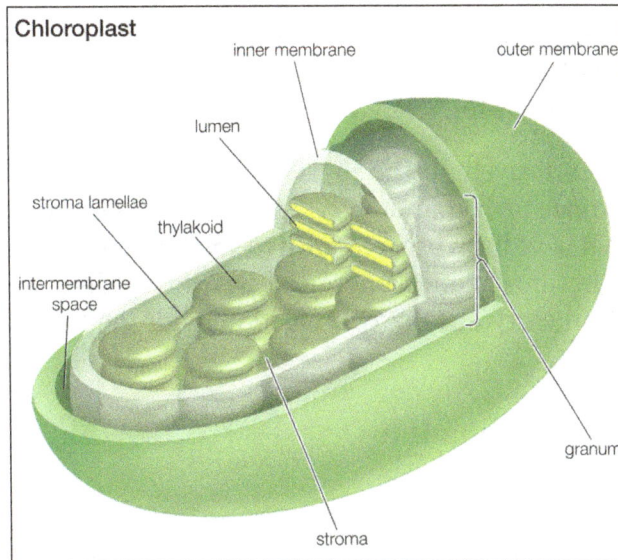

Chloroplast structure.

The internal (thylakoid) membrane vesicles are organized into stacks, which reside in a matrix known as the stroma. All the chlorophyll in the chloroplast is contained in the membranes of the thylakoid vesicles.

Structural Features

The intricate structural organization of the photosynthetic apparatus is essential for the efficient performance of the complex process of photosynthesis. The chloroplast is enclosed in a double outer membrane, and its size approximates a spheroid about 2,500 nm thick and 5,000 nm long. Some single-celled algae have one chloroplast that occupies more than half the cell volume. Leaf cells of higher plants contain many chloroplasts, each approximately the size of the one in some algal cells.

When thin sections of a chloroplast are examined under the electron microscope, several features are apparent. Chief among these are the intricate internal membranes (i.e., the lamellae) and the stroma, a colourless matrix in which the lamellae are embedded. Also visible are starch granules, which appear as dense bodies.

The stroma is basically a solution of enzymes and small molecules. The dark reactions occur in the stroma, the soluble enzymes of which catalyze the conversion of carbon dioxide and minerals to carbohydrates and other organic compounds. The capacity for carbon fixation and reduction is lost if the outer membrane of the chloroplast is broken, allowing the stroma enzymes to leak out.

A single lamella, which contains all the photosynthetic pigments, is approximately 10–15 nm thick. The lamellae exist in more-or-less flat sheets, a few of which extend through much of the length of the chloroplast. Examination of cross sections of lamellae under the electron microscope shows that their edges are joined to form closed hollow disks that are called thylakoids ("saclike"). The chloroplasts of most higher plants have regions, called grana, in which the thylakoids are very tightly stacked. When viewed by electron microscopy at an oblique angle, the grana appear as stacks of disks. When viewed in cross section, it is apparent that some thylakoids extend from one grana through the stroma into other grana. The thin aqueous spaces inside the thylakoids are believed to be connected with each other via these stroma thylakoids. These thylakoid spaces are isolated from the stroma spaces by the relatively impermeable lamellae.

The light reactions occur exclusively in the thylakoids. The complex structural organization of lamellae is required for proper thylakoid function; intact thylakoids are necessary for the formation of ATP. Thylakoids that have been broken down to smaller units can no longer form ATP, even when the conversion of light into chemical energy occurs during electron transport in these units. Such lamellar fragments can carry out the Hill reaction, with the transfer of electrons from water to $NADP^+$.

Chemical Composition of Lamellae

Lipids

Lamellae consist of about equal amounts of lipids and proteins. About one-fourth of the lipid portion of the lamellae consists of pigments and coenzymes; the remainder consists of various lipids, including polar compounds such as phospholipids and galactolipids. These polar lipid molecules have "head" groups that attract water (i.e., are hydrophilic) and fatty acid "tails" that are oil soluble and repel water (i.e., are hydrophobic). When polar lipids are placed in an aqueous environment, they can line up with the fatty acid tails side by side. A second layer of phospholipids forms tail-to-tail with the first, establishing a lipid bilayer in which the hydrophilic heads are in contact with the aqueous solution on each side of the bilayer. Sandwiched between the heads are the hydrophobic tails, creating a hydrophobic environment from which water is excluded. This lipid bilayer is an essential feature of all biological membranes. The hydrophobic parts of proteins and lipid-soluble cofactors and pigments are dissolved or embedded in the lipid bilayer. Lamellar membranes can function as electrical insulating material and permit a charge, or potential difference, to develop across the membrane. Such a charge can be a source of chemical or electrical energy.

Approximately one-fifth of the lamellar lipids are chlorophyll molecules; one type, chlorophyll a, is more abundant than the second type, chlorophyll b. The chlorophyll molecules are specifically bound to small protein molecules. Most of these chlorophyll-proteins are "light-harvesting" pigments. These absorb light and pass its energy on to special chlorophyll a molecules that are directly involved in the conversion of light energy to chemical energy. When one of these special chlorophyll a molecules is excited by light energy (as described later), it gives up an electron. There are two types of these special chlorophyll a molecules: one, called P_{680}, has an absorption spectrum that peaks at 684 nm; the other, called P_{700}, shows an absorption peak at 700 nm.

Although chlorophylls are the main light-absorbing molecules in green plants, there are other pigments such as carotenes and carotenoids (which are responsible for the yellow-orange colour of carrots). Carotenes can also absorb light and may supplement chlorophyll as the light-absorbing molecules in some plant cells. The light energy absorbed by carotenes must be passed to chlorophyll before conversion to chemical energy can occur. Carotenoids are part of a cycle that renders excess energy beyond the level of light saturation harmless, effectively serving as "lightning rods" in the process.

Proteins

Many of the lamellar proteins are components of the chlorophyll–protein complexes described above. Other proteins include enzymes and protein-containing coenzymes. Enzymes are required as organic catalysts for specific reactions within the lamellae. Protein coenzymes, also called cofactors, include important electron carrier molecules called cytochromes, which are iron-containing pigments with the pigment portions attached to protein molecules. During electron transfer, an electron is accepted by an iron atom in the pigment portion of a cytochrome molecule, which thus is reduced; then the electron is transferred to the iron atom in the next cytochrome carrier in the electron transfer chain, thus oxidizing the first cytochrome and reducing the next one in the chain.

In addition to the metal atoms found in the pigment portions of cytochrome molecules, metal atoms also are found in other protein molecules of the lamellae. In proteins with a total molecular weight of 900,000 (based on the weight of hydrogen as one), there are 2 atoms of manganese, 10 atoms of iron, and 6 atoms of copper. These metal atoms are required for the catalytic activity of some of the enzymes important in photosynthesis. The manganese atoms are involved in water-splitting and oxygen formation. Both copper- and iron-containing proteins function in electron transport between water and the final electron-acceptor molecule of the light stage of photosynthesis, an iron-containing protein called ferredoxin. Ferredoxin is a soluble component in the chloroplasts. In its reduced form, it gives electrons directly to the systems that reduce nitrate and sulfate and via NADPH to the system that reduces carbon dioxide. A copper-containing protein called plastocyanin (PC) carries electrons at one point in the electron transport chain. PC molecules are water soluble and can move through the inner space of the thylakoids, carrying electrons from one place to another.

Quinones

Small molecules called plastoquinones are found in substantial numbers in the lamellae. Like the cytochromes, quinones have important roles in carrying electrons between the components of the light reactions. Since they are lipid soluble, they can diffuse through the membrane. They can carry one or two electrons, and, in their reduced form (with added electrons), they carry hydrogen atoms that can be released as hydrogen ions when the added electrons are passed on, for example, to a cytochrome.

The Process of Photosynthesis: The Light Reactions

Light Absorption and Energy Transfer

The light energy absorbed by a chlorophyll molecule excites some electrons within the structure of the molecule to higher energy levels, or excited states. Light of shorter wavelength (such as blue) has more energy than light of longer wavelength (such as red), so absorption of blue light creates an excited state of higher energy. A molecule raised to this higher energy state quickly gives up the "extra" energy as heat and falls to its lowest excited state. This lowest excited state is similar to that of a molecule that has just absorbed the longest wavelength light capable of exciting it. In the case of chlorophyll *a*, this lowest excited state corresponds to that of a molecule that has absorbed red light of about 680 nm.

The return of a chlorophyll *a* molecule from its lowest excited state to its original low-energy state (ground state) requires the release of the extra energy of the excited state. This can occur in one of several ways. In photosynthesis, most of this energy is conserved as chemical energy by the transfer of an electron from a special chlorophyll *a* molecule (P_{680} or P_{700}) to an electron acceptor. When this electron transfer is blocked by inhibitors, such as the herbicide dichlorophenylmethylurea (DCMU), or by low temperature, the energy can be released as red light. Such reemission of light is called fluorescence. The examination of fluorescence from photosynthetic material in which electron transfer has been blocked has proved to be a valuable tool for scientists studying the light reactions.

The Pathway of Electrons

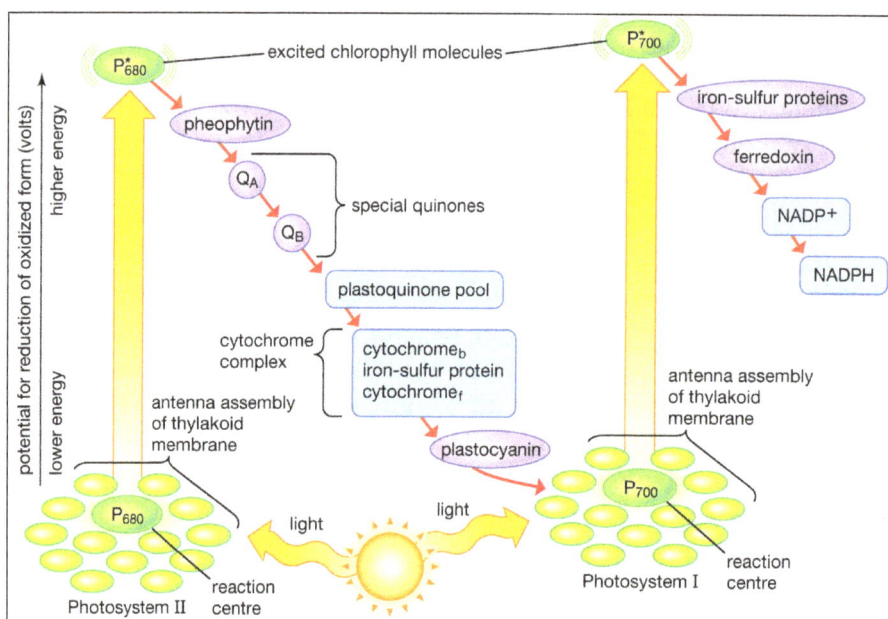

Flow of electrons during the light reaction stage of photosynthesis Arrows pointing upward represent light reactions that increase the chemical potential; arrows slanting downward represent flow of electrons via carriers in the membrane.

The general features of a widely accepted mechanism for photoelectron transfer, in which two light reactions (light reaction I and light reaction II) occur during the transfer of electrons from water to carbon dioxide, were proposed by Robert Hill and Fay Bendall in 1960. This mechanism is based on the relative potential (in volts) of various cofactors of the electron-transfer chain to be oxidized or reduced. Molecules that in their oxidized form have the strongest affinity for electrons (i.e., are strong oxidizing agents) have a low relative potential. In contrast, molecules that in their oxidized form are difficult to reduce have a high relative potential once they have accepted electrons. The molecules with a low relative potential are considered to be strong oxidizing agents, and those with a high relative potential are considered to be strong reducing agents.

In diagrams that describe the light reaction stage of photosynthesis, the actual photochemical steps are typically represented by two vertical arrows. These arrows signify that the special pigments P_{680} and P_{700} receive light energy from the light-harvesting chlorophyll-protein molecules and are raised in energy from their ground state to excited states. In their excited state, these pigments are extremely strong reducing agents that quickly transfer electrons to the first acceptor. These first acceptors also are strong reducing agents and rapidly pass electrons to more stable carriers. In light reaction II, the first acceptor may be pheophytin, which is a molecule similar to chlorophyll that also has a strong reducing potential and quickly transfers electrons to the next acceptor. Special quinones are next in the series. These molecules are similar to plastoquinone; they receive electrons from pheophytin and pass them to the intermediate electron carriers, which include the plastoquinone pool and the cytochromes b and f associated in a complex with an iron-sulfur protein.

In light reaction I, electrons are passed on to iron-sulfur proteins in the lamellar membrane, after which the electrons flow to ferredoxin, a small water-soluble iron-sulfur protein. When $NADP^+$ and a suitable enzyme are present, two ferredoxin molecules, carrying one electron each, transfer two electrons to $NADP^+$, which picks up a proton (i.e., a hydrogen ion) and becomes NADPH.

Each time a P_{680} or P_{700} molecule gives up an electron, it returns to its ground (unexcited) state, but with a positive charge due to the loss of the electron. These positively charged ions are extremely strong oxidizing agents that remove an electron from a suitable donor. The P_{680}^+ of light reaction II is capable of taking electrons from water in the presence of appropriate catalysts. There is good evidence that two or more manganese atoms complexed with protein are involved in this catalysis, taking four electrons from two water molecules (with release of four hydrogen ions). The manganese-protein complex gives up these electrons one at a time via an unidentified carrier to P_{680}^+, reducing it to P_{680}. When manganese is selectively removed by chemical treatment, the thylakoids lose the capacity to oxidize water, but all other parts of the electron pathway remain intact.

In light reaction I, P_{700}^+ recovers electrons from plastocyanin, which in turn receives them from intermediate carriers, including the plastoquinone pool and cytochrome b and cytochrome f molecules. The pool of intermediate carriers may receive electrons from water via light reaction II and the quinones. Transfer of electrons from water to ferredoxin via the two light reactions and intermediate carriers is called noncyclic electron flow. Alternatively, electrons may be transferred only by light reaction I, in which case they are recycled from ferredoxin back to the intermediate carriers. This process is called cyclic electron flow.

Evidence of Two Light Reactions

Many lines of evidence support the concept of electron flow via two light reactions. An early study by American biochemist Robert Emerson employed the algae *Chlorella*, which was illuminated with red light alone, with blue light alone, and with red and blue light at the same time. Oxygen evolution was measured in each case. It was substantial with blue light alone but not with red light alone. With both red and blue light together, the amount of oxygen evolved far exceeded the sum of that seen with blue and red light alone. These experimental data pointed to the existence of two types of light reactions that, when operating in tandem, would yield the highest rate of oxygen evolution. It is now known that light reaction I can use light of a slightly longer wavelength than red ($\lambda = 680$ nm), while light reaction II requires light with a wavelength of 680 nm or shorter.

Since those early studies, the two light reactions have been separated in many ways, including separation of the membrane particles in which each reaction occurs. As discussed previously, lamellae can be disrupted mechanically into fragments that absorb light energy and break the bonds of water molecules (i.e., oxidize water) to produce oxygen, hydrogen ions, and electrons. These electrons can be transferred to ferredoxin, the final electron acceptor of the light stage. No transfer of electrons from water to ferredoxin occurs if the herbicide DCMU is present. The subsequent addition of certain reduced dyes (i.e., electron donors) restores the light reduction of $NADP^+$ but without oxygen production, suggesting that light reaction I but not light reaction II is functioning. It is now known that DCMU blocks the transfer of electrons between the first quinone and the plastoquinone pool in light reaction II.

When treated with certain detergents, lamellae can be broken down into smaller particles capable of carrying out single light reactions. One type of particle can absorb light energy, oxidize water, and produce oxygen (light reaction II), but a special dye molecule must be supplied to accept the electrons. In the presence of electron donors, such as a reduced dye, a second type of lamellar particle can absorb light and transfer electrons from the electron donor to ferredoxin (light reaction I).

Photosystems I and II

The structural and photochemical properties of the minimum particles capable of performing light reactions I and II have received much study. Treatment of lamellar fragments with neutral detergents releases these particles, designated photosystem I and photosystem II, respectively. Subsequent harsher treatment (with charged detergents) and separation of the individual polypeptides with electrophoretic techniques have helped identify the components of the photosystems. Each photosystem consists of a light-harvesting complex and a core complex. Each core complex contains a reaction centre with the pigment (either P_{700} or P_{680}) that can be photochemically oxidized, together with electron acceptors and electron donors. In addition, the core complex has some 40 to 60 chlorophyll molecules bound to proteins. In addition to the light absorbed by the chlorophyll molecules in the core complex, the reaction centres receive a major part of their excitation from the pigments of the light-harvesting complex.

Quantum Requirements

The quantum requirements of the individual light reactions of photosynthesis are defined as the number of light photons absorbed for the transfer of one electron. The quantum requirement for each light reaction has been found to be approximately one photon. The total number of quanta required, therefore, to transfer the four electrons that result in the formation of one molecule of oxygen via the two light reactions should be four times two, or eight. It appears, however, that additional light is absorbed and used to form ATP by a cyclic photophosphorylation pathway. (The cyclic photophosphorylation pathway is an ATP-forming process in which the excited electron returns to the reaction centre.) The actual quantum requirement, therefore, probably is 9 to 10.

The Process of Photosynthesis: The Conversion of Light Energy to ATP

The electron transfers of the light reactions provide the energy for the synthesis of two compounds vital to the dark reactions: NADPH and ATP. The previous section explained how noncyclic electron flow results in the reduction of $NADP^+$ to NADPH. In this section, the synthesis of the energy-rich compound ATP is described.

ATP is formed by the addition of a phosphate group to a molecule of adenosine diphosphate (ADP)—or to state it in chemical terms, by the phosphorylation of ADP. This reaction requires a substantial input of energy, much of which is captured in the bond that links the added phosphate group to ADP. Because light energy powers this reaction in the chloroplasts, the production of ATP during photosynthesis is referred to as photophosphorylation, as opposed to oxidative phosphorylation in the electron-transport chain in the mitochondrion.

Unlike the production of NADPH, the photophosphorylation of ADP occurs in conjunction with both cyclic and noncyclic electron flow. In fact, researchers speculate that the sole purpose of cyclic electron flow may be for photophosphorylation, since this process involves no net transfer of electrons to reducing agents. The relative amounts of cyclic and noncyclic flow may be adjusted in accordance with changing physiological needs for ATP and reduced ferredoxin and NADPH in chloroplasts. In contrast to electron transfer in light reactions I and II, which can occur in membrane fragments, intact thylakoids are required for efficient photophosphorylation. This requirement stems from the special nature of the mechanism linking photophosphorylation to electron flow in the lamellae.

The theory relating the formation of ATP to electron flow in the membranes of both chloroplasts and mitochondria (the organelles responsible for ATP formation during cellular respiration) was first proposed by English biochemist Peter Dennis Mitchell, who received the 1978 Nobel Prize for Chemistry. This chemiosmotic theory has been somewhat modified to fit later experimental facts. The general features are now widely accepted. A central feature is the formation of a hydrogen ion (proton) concentration gradient and an electrical charge across intact lamellae. The potential energy stored by the proton gradient and electrical charge is then used to drive the energetically unfavourable conversion of ADP and inorganic phosphate (P_i) to ATP and water.

The manganese-protein complex associated with light reaction II is exposed to the interior of the thylakoid. Consequently, the oxidation of water during light reaction II leads to release of hydrogen ions (protons) into the inner thylakoid space. Furthermore, it is likely that photoreaction II entails the transfer of electrons across the lamella toward its outer face, so that when plastoquinone molecules are reduced, they can receive protons from the outside of the thylakoid. When these reduced plastoquinone molecules are oxidized, giving up electrons to the cytochrome-iron-sulfur complex, protons are released inside the thylakoid. Because the lamella is impermeable to them, the release of protons inside the thylakoid by oxidation of both water and plastoquinone leads to a higher concentration of protons inside the thylakoid than outside it. In other words, a proton gradient is established across the lamella. Since protons are positively charged, the movement of protons across the thylakoid lamella during both light reactions results in the establishment of an electrical charge across the lamella.

An enzyme complex located partly in and on the lamellae catalyzes the reaction in which ATP is formed from ADP and inorganic phosphate. The reverse of this reaction is catalyzed by an enzyme called ATP-ase; hence, the enzyme complex is sometimes called an ATP-ase complex. It is also called the coupling factor. It consists of hydrophilic polypeptides (F_1), which project from the outer surface of the lamellae, and hydrophobic polypeptides (F_0), which are embedded inside the lamellae. F_0 forms a channel that permits protons to flow through the lamellar membrane to F_1. The enzymes in F_1 then catalyze ATP formation, using both the proton supply and the lamellar transmembrane charge.

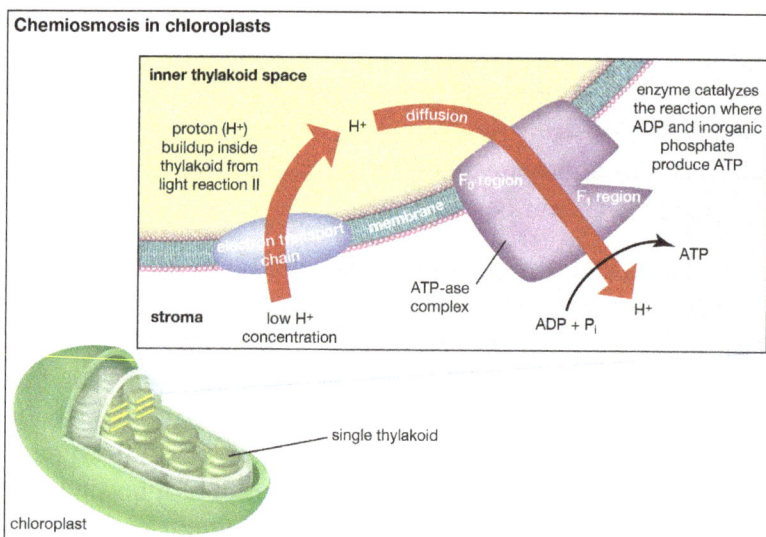

Chemiosmosis in chloroplasts

Chemiosmosis in chloroplasts that results in the donation of a
proton for the production of adenosine triphosphate (ATP) in plants.

In summary, the use of light energy for ATP formation occurs indirectly: a proton gradient and electrical charge—built up in or across the lamellae as a consequence of electron flow in the light reactions—provide the energy to drive the synthesis of ATP from ADP and P_i.

The Process of Photosynthesis: Carbon Fixation and Reduction

The assimilation of carbon into organic compounds is the result of a complex series of enzymatically regulated chemical reactions—the dark reactions. This term is something of a misnomer, for these reactions can take place in either light or darkness. Furthermore, some of the enzymes involved in the so-called dark reactions become inactive in prolonged darkness; however, they are activated when the leaves that contain them are exposed to light.

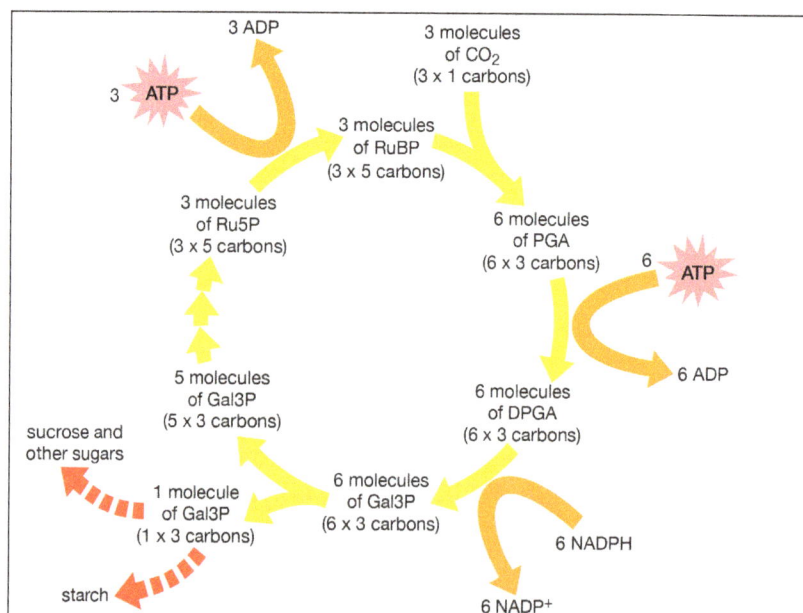

C_3 carbon fixation pathway.

Pathway of carbon dioxide fixation and reduction in photosynthesis, the reductive pentose phosphate cycle. The diagram represents one complete turn of the cycle, with the net production of one molecule of Gal3P. The nine molecules of ATP and six molecules of NADPH come from the light reactions.

Elucidation of the Carbon Pathway

Radioactive isotopes of carbon (^{14}C) and phosphorus (^{32}P) have been valuable in identifying the intermediate compounds formed during carbon assimilation. A photosynthesizing plant does not strongly discriminate between the most abundant natural carbon isotope (^{12}C) and ^{14}C. During photosynthesis in the presence of $^{14}CO_2$, the compounds formed become labeled with the radioisotope. During very short exposures, only the first intermediates in the carbon-fixing pathway become labeled. Early investigations showed that some radioactive products were formed even when the light was turned off and the $^{14}CO_2$ was added just afterward in the dark, confirming the nature of the carbon fixation as a "dark" reaction.

American biochemist Melvin Calvin, a Nobel Prize recipient for his work on the carbon-reduction cycle, allowed green plants to photosynthesize in the presence of radioactive carbon dioxide for a few seconds under various experimental conditions. Products that became labeled with radioactive carbon during Calvin's experiments included a three-carbon compound called 3-phosphoglycerate (abbreviated PGA), sugar phosphates, amino acids, sucrose, and carboxylic acids. When photosynthesis was stopped after two seconds, the principal radioactive product was PGA, which therefore was identified as the first stable compound formed during carbon dioxide fixation in green plants. PGA is a three-carbon compound, and the mode of photosynthesis is thus referred to as C_3. In the two other known pathways, C_4 and CAM (crassulacean acid metabolism), the C_3 pathway follows the fixation of CO_2 into oxaloacetate, a four-carbon acid, and its reduction to malate. PGA is formed from 2-carboxy-3-keto-D-arabinitol 1,5-bisphosphate, which is a highly unstable six-carbon compound formed from the carboxylation of ribulose-1,5-bisphosphate, a five-carbon compound.

Further studies with ^{14}C as well as with inorganic phosphate labeled with ^{32}P led to the mapping of the carbon fixation and reduction pathway called the reductive pentose phosphate (RPP) cycle, or the Calvin-Benson cycle. An additional pathway for carbon transport in certain plants was later discovered in other laboratories. All the steps in these pathways can be carried out in the laboratory by isolated enzymes in the dark. Several steps require the ATP or NADPH generated by the light reactions. In addition, some of the enzymes are fully active only when conditions simulate those in green cells exposed to light. In living plants, these enzymes are active during photosynthesis but not in the dark.

The Calvin-Benson Cycle

The Calvin-Benson cycle, in which carbon is fixed, reduced, and utilized, involves the formation of intermediate sugar phosphates in a cyclic sequence. One complete cycle incorporates three molecules of carbon dioxide and produces one molecule of the three-carbon compound glyceraldehyde-3-phosphate (Gal3P). This three-carbon sugar phosphate usually is either exported from the chloroplasts or converted to starch inside the chloroplast.

ATP and NADPH formed during the light reactions are utilized for key steps in this pathway and provide the energy and reducing equivalents (i.e., electrons) to drive the sequence in the direction shown. For each molecule of carbon dioxide that is fixed, two molecules of NADPH and three molecules of ATP from the light reactions are required. The overall reaction can be represented as follows:

$$9\,ATP + 6\,NADPH + 3\,CO_2 \rightarrow Gal3P + 6\,NADP^+ + 9\,ADP + 8\,P_i$$

The cycle is composed of four stages: (1) carboxylation, (2) reduction, (3) isomerization/condensation/dismutation, and (4) phosphorylation.

Carboxylation

The initial incorporation of carbon dioxide, which is catalyzed by the enzyme ribulose 1,5-bisphosphate carboxylase (Rubisco), proceeds by the addition of carbon dioxide to the five-carbon compound ribulose 1,5-bisphosphate (RuBP) and the splitting of the resulting six-carbon compound into two molecules of PGA. This reaction occurs three times during each complete turn of the cycle; thus, six molecules of PGA are produced.

Reduction

The six molecules of PGA are first phosphorylated with ATP by the enzyme PGA-kinase, yielding six molecules of 1,3-diphosphoglycerate (DPGA). These molecules are subsequently reduced with NADPH and the enzyme glyceraldehyde-3-phosphate dehydrogenase to give six molecules of Gal3P. These reactions are the reverse of two steps of the process glycolysis in cellular respiration.

Isomerization/Condensation/Dismutation

For each complete Calvin-Benson cycle, one of the Gal3P molecules, with its three carbon atoms, is the net product and may be transferred out of the chloroplast or converted to starch inside the chloroplast. For the cycle to regenerate, the other five Gal3P molecules (with a total of 15 carbon atoms) must be converted back to three molecules of five-carbon RuBP. The conversion of Gal3P to RuBP begins with a complex series of enzymatically regulated reactions that lead to the synthesis of the five-carbon compound ribulose-5-phosphate (Ru5P).

Phosphorylation

The three molecules of Ru5P are converted to the carboxylation substrate, RuBP, by the enzyme phosphoribulokinase, using ATP. This reaction, shown below, completes the cycle:

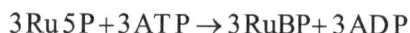

$$3\,Ru5P + 3\,ATP \rightarrow 3\,RuBP + 3\,ADP$$

Regulation of the Cycle

Photosynthesis cannot occur at night, but the respiratory process of glycolysis—which uses some of the same reactions as the Calvin-Benson cycle, except in the reverse—does take place. Thus, some steps in this cycle would be wasteful if allowed to occur in the dark, because they would counteract the reactions of glycolysis. For this reason, some enzymes of the Calvin-Benson cycle are "turned off" (i.e., become inactive) in the dark.

Even in the presence of light, changes in physiological conditions frequently necessitate adjustments in the relative rates of reactions of the Calvin-Benson cycle, so that enzymes for some reactions change in their catalytic activity. These alterations in enzyme activity typically are brought about by changes in levels of such chloroplast components as reduced ferredoxin, acids, and soluble components (e.g., P_i and magnesium ions).

Products of Carbon Reduction

The most important use of Gal3P is its export from the chloroplasts to the cytosol of green cells, where it is used for biosynthesis of products needed by the plant. In land plants, a principal product is sucrose, which is translocated from the green cells of the leaves to other parts of the plant. Other key products include the carbon skeletons of certain primary amino acids, such as alanine, glutamate, and aspartate. To complete the synthesis of these compounds, amino groups are added to the appropriate carbon skeletons made from Gal3P. Sulfur amino acids such as cysteine are formed by adding sulfhydryl groups and amino groups. Other biosynthesis pathways lead from Gal3P to lipids, pigments, and most of the constituents of green cells.

Starch synthesis and accumulation in the chloroplasts occur particularly when photosynthetic carbon fixation exceeds the needs of the plant. Under such circumstances, sugar phosphates accumulate in the cytosol, binding cytosolic P_i. The export of Gal3P from the chloroplasts is tied to a one-for-one exchange of P_i for Gal3P, so less cytosolic P_i results in decreased export of Gal3P and decreased P_i in the chloroplast. These changes trigger alterations in the activities of regulated enzymes, leading in turn to increased starch synthesis. This starch can be broken down at night and used as a source of reduced carbon and energy for the physiological needs of the plant. Too much starch in the chloroplasts leads to diminished rates of photosynthesis, however. In addition, high levels of sugars in the cytosol lead to the suppression of the normal activities of the genes involved in photosynthesis. Thus, under what would seem to be the ideal photosynthetic conditions of a bright warm day, many plants in fact have-slower-than expected rates of photosynthesis.

Photorespiration

Under conditions of high light intensity, hot weather, and water limitation, the productivity of the Calvin-Benson cycle is limited in many plants by the occurrence of photorespiration. This process converts sugar phosphates back to carbon dioxide; it is initiated by the oxygenation of RuBP (i.e., the combination of gaseous oxygen [O_2] with RuBP). This oxygenation reaction yields only one molecule of PGA and one molecule of a two-carbon acid, phosphoglycolate, which is subsequently converted in part to carbon dioxide. The reaction of oxygen with RuBP is in direct competition with the carboxylation reaction (CO_2 + RuBP) that initiates the Calvin-Benson cycle and is, in fact, catalyzed by the same protein, ribulose 1,5-bisphosphate carboxylase. The relative concentrations of oxygen and carbon dioxide within the chloroplasts as well as leaf temperature determine whether oxygenation or carboxylation is favoured. The concentration of oxygen inside the chloroplasts may be higher than atmospheric (20 percent) because of photosynthetic oxygen evolution, whereas the internal carbon dioxide concentration may be lower than atmospheric (0.039 percent) because of photosynthetic uptake. Any increase in the internal carbon dioxide pressure tends to help the carboxylation reaction compete more effectively with oxygenation.

Carbon Fixation in C_4 Plants

Certain plants—including the important crops sugarcane and corn (maize), as well as other diverse species that are thought to have expanded their geographic ranges into tropical areas—have developed a special mechanism of carbon fixation that largely prevents photorespiration. The leaves of these plants have special anatomy and biochemistry. In particular, photosynthetic functions are divided between mesophyll and bundle-sheath leaf cells. The carbon-fixation pathway begins in the mesophyll cells, where carbon dioxide is converted into bicarbonate, which is then added to the three-carbon acid phosphoenolpyruvate (PEP) by an enzyme called phosphoenolpyruvate carboxylase. The product of this reaction is the four-carbon acid oxaloacetate, which is reduced to malate, another four-carbon acid, in one form of the C_4 pathway. Malate then is transported to bundle-sheath cells, which are located near the vascular system of the leaf. There, malate enters the chloroplasts and is oxidized and decarboxylated (i.e., loses CO_2) by malic enzyme. This yields high concentrations of carbon dioxide, which is fed into the Calvin-Benson cycle of the bundle sheath cells, and pyruvate, a three-carbon acid that is translocated back to the mesophyll cells. In the mesophyll chloroplasts, the enzyme pyruvate orthophosphate dikinase (PPDK) uses ATP and P_i to convert pyruvate back to PEP, completing the C_4 cycle. There are several variations of this pathway in different species. For example, the amino acids aspartate and alanine can substitute for malate and pyruvate in some species.

The C_4 pathway acts as a mechanism to build up high concentrations of carbon dioxide in the chloroplasts of the bundle sheath cells. The resulting higher level of internal carbon dioxide in these chloroplasts serves to increase the ratio of carboxylation to oxygenation, thus minimizing photorespiration. Although the plant must expend extra energy to drive this mechanism, the energy loss is more than compensated by the near elimination of photorespiration under conditions where it would otherwise occur. Sugarcane and certain other plants that employ this pathway have the highest annual yields of biomass of all species. In cool climates, where photorespiration is insignificant, C_4 plants are rare. Carbon dioxide is also used efficiently in carbohydrate synthesis in the bundle sheath.

PEP carboxylase, which is located in the mesophyll cells, is an essential enzyme in C_4 plants. In hot and dry environments, carbon dioxide concentrations inside the leaf fall when the plant closes or partially closes its stomata to reduce water loss from the leaves. Under these conditions, photorespiration is likely to occur in plants that use Rubisco as the primary carboxylating enzyme, since Rubisco adds oxygen to RuBP when carbon dioxide concentrations are low. PEP carboxylase, however, does not use oxygen as a substrate, and it has a greater affinity for carbon dioxide than Rubisco does. Thus, it has the ability to fix carbon dioxide in reduced carbon dioxide conditions, such as when the stomata on the leaves are only partially open. As a consequence, at similar rates of photosynthesis, C_4 plants lose less water when compared with C_3 plants. This explains why C_4 plants are favoured in dry and warm environments.

Carbon Fixation via Crassulacean Acid Metabolism (CAM)

In addition to C_3 and C_4 species, there are many succulent plants that make use of a third photosynthetic pathway: crassulacean acid metabolism (CAM). This pathway is named after the Crassulaceae, a family in which many species display this type of metabolism, but it also occurs commonly in other families, such as the Cactaceae, the Euphorbiaceae, the Orchidaceae, and the Bromeliaceae. CAM species number more than 20,000 and span 34 families. Almost all CAM plants are angiosperms; however, quillworts and ferns also use the CAM pathway. In addition, some scientists note that CAM might be used by *Welwitschia*, a gymnosperm. CAM plants are often characterized by their succulence, but this quality is not pronounced in epiphytes that use the CAM pathway.

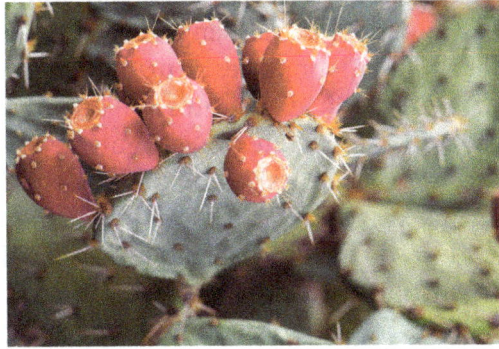

Prickly Pear Prickly pear cactus (Opuntia), Arizona, U.S.

CAM plants are known for their capacity to fix carbon dioxide at night, using PEP carboxylase as the primary carboxylating enzyme and the accumulation of malate (which is made by the enzyme malate dehydrogenase) in the large vacuoles of their cells. Deacidification occurs during the day, when carbon dioxide is released from malate and fixed in the Calvin-Benson cycle, using Rubisco. During daylight hours, the stomata are closed to prevent water loss. The stomata are open at night when the air is cooler and more humid, and this setting allows the leaves of the plant to assimilate carbon dioxide. Since their stomata are closed during the day, CAM plants require considerably less water than both C_3 and C_4 plants that fix the same amount of carbon dioxide in photosynthesis.

The productivity of most CAM plants is fairly low, however. This is not an inherent trait of CAM species, because some cultivated CAM plants (e.g., *Agave mapisaga* and *A. salmiana*) can achieve a high aboveground productivity. In fact, some cultivated species that are irrigated, fertilized, and carefully pruned are highly productive. For example, prickly pear (*Opuntia ficus-indica*) and its thornless variety, *O. amyclea*, produce 4.6 kg per square metre (0.9 pound per square foot) of new growth per year. Such productivity is among the highest of any plant species. Thus, the rates of photosynthesis of CAM plants may be as high as those of C_3 plants, if morphologically similar plants adapted to the similar habitats are compared.

The unusual capacity of CAM plants to fix carbon dioxide into organic acids in the dark, causing nocturnal acidification, with deacidification occurring during the day, has been known to science since the 19th century. (There is evidence, however, that the Romans noticed the difference between the morning acid taste of some of the house plants they cultivated.) On the other hand, the C_4 pathway was discovered during the middle of the 20th century. A full appreciation of CAM as a photosynthetic pathway was greatly stimulated by analogies with C_4 species.

The Molecular Biology of Photosynthesis

Oxygenic photosynthesis occurs in a certain type of prokaryotic cells called cyanobacteria and eukaryotic plant cells (algae and higher plants). In eukaryotic plant cells, which contain chloroplasts and a nucleus, the genetic information needed for the reproduction of the photosynthetic apparatus is contained partly in the chloroplast chromosome and partly in chromosomes of the nucleus. For example, the carboxylation enzyme ribulose 1,5-bisphosphate carboxylase is a large protein molecule comprising a complex of eight large polypeptide subunits and eight small polypeptide subunits. The gene for the large subunits is located in the chloroplast chromosome, whereas the gene for the small subunits is in the nucleus. Transcription of the DNA of the nuclear gene yields messenger RNA (mRNA) that encodes the information for the synthesis of the small polypeptides. During this synthesis, which occurs on the cytosolic ribosomes, some extra amino acid residues are added to form a recognition leader on the end of the polypeptide chain. This leader is recognized by special receptor sites on the outer chloroplast membrane; these receptor sites then allow the polypeptide to penetrate the membrane and enter the chloroplast. The leader is removed, and the small subunits combine with the large subunits, which have been synthesized on chloroplast ribosomes according to mRNA transcribed from the chloroplast DNA. The expression of nuclear genes that code for proteins needed in the chloroplasts appears to be under control of events in the chloroplasts in some cases; for example, the synthesis of some nuclear-encoded chloroplast enzymes may occur only when light is absorbed by chloroplasts.

Artificial Photosynthesis

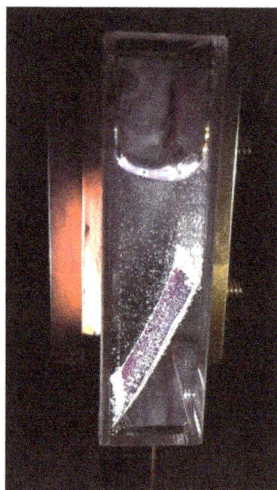

A sample of a photoelectric cell in a lab environment. Catalysts are added to the cell, which is submerged in water and illuminated by simulated sunlight. The bubbles seen are oxygen (forming on the front of the cell) and hydrogen (forming on the back of the cell).

Artificial photosynthesis is a chemical process that biomimics the natural process of photosynthesis to convert sunlight, water, and carbon dioxide into carbohydrates and oxygen. The term artificial photosynthesis is commonly used to refer to any scheme for capturing and storing the energy from sunlight in the chemical bonds of a fuel (a solar fuel). Photocatalytic water splitting converts water into hydrogen and oxygen and is a major research topic of artificial photosynthesis. Light-driven carbon dioxide reduction is another process studied that replicates natural carbon fixation.

Research of this topic includes the design and assembly of devices for the direct production of solar fuels, photoelectrochemistry and its application in fuel cells, and the engineering of enzymes and photoautotrophic microorganisms for microbial biofuel and biohydrogen production from sunlight.

The photosynthetic reaction can be divided into two half-reactions of oxidation and reduction, both of which are essential to producing fuel. In plant photosynthesis, water molecules are photo-oxidized to release oxygen and protons. The second phase of plant photosynthesis (also known as the Calvin-Benson cycle) is a light-independent reaction that converts carbon dioxide into glucose (fuel). Researchers of artificial photosynthesis are developing photocatalysts that are able to perform both of these reactions. Furthermore, the protons resulting from water splitting can be used for hydrogen production. These catalysts must be able to react quickly and absorb a large percentage of the incident solar photons.

Natural (Left) vs. Artificial Photosynthesis (Right)

Whereas photovoltaics can provide energy directly from sunlight, the inefficiency of fuel production from photovoltaic electricity (indirect process) and the fact that sunshine is not constant throughout the day sets a limit to its use. One way of using natural photosynthesis is for the production of a biofuel, which is an indirect process that suffers from low energy conversion efficiency (due to photosynthesis' own low efficiency in converting sunlight to biomass), the cost of harvesting and transporting the fuel, and conflicts due to the increasing need of land mass for food production. The purpose of artificial photosynthesis is to produce a fuel from sunlight that can be stored conveniently and used when sunlight is not available, by using direct processes, that is, to produce a solar fuel. With the development of catalysts able to reproduce the major parts of photosynthesis, water and sunlight would ultimately be the only needed sources for clean energy production. The only by-product would be oxygen, and production of a solar fuel has the potential to be cheaper than gasoline.

One process for the creation of a clean and affordable energy supply is the development of photocatalytic water splitting under solar light. This method of sustainable hydrogen production is a major objective for the development of alternative energy systems. It is also predicted to be one of the more, if not the most, efficient ways of obtaining hydrogen from water. The conversion of solar energy into hydrogen via a water-splitting process assisted by photosemiconductor catalysts is one of the most promising technologies in development. This process has the potential for large quantities of hydrogen to be generated in an ecologically sound manner. The conversion of solar energy into a clean fuel (H_2) under ambient conditions is one of the greatest challenges facing scientists in the twenty-first century.

Two methods are generally recognized for the construction of solar fuel cells for hydrogen production:

- A homogeneous system is one such that catalysts are not compartmentalized, that is, components are present in the same compartment. This means that hydrogen and oxygen are produced in the same location. This can be a drawback, since they compose an explosive mixture, demanding gas product separation. Also, all components must be active in approximately the same conditions (e.g., pH).

- A heterogeneous system has two separate electrodes, an anode and a cathode, making possible the separation of oxygen and hydrogen production. Furthermore, different components do not necessarily need to work in the same conditions. However, the increased complexity of these systems makes them harder to develop and more expensive.

Another area of research within artificial photosynthesis is the selection and manipulation of photosynthetic microorganisms, namely green microalgae and cyanobacteria, for the production of solar fuels. Many strains are able to produce hydrogen naturally, and scientists are working to improve them. Algae biofuels such as butanol and methanol are produced both at laboratory and commercial scales. This method has benefited from the development of synthetic biology, which is also being explored by the J. Craig Venter Institute to produce a synthetic organism capable of biofuel production. In 2017, an efficient process was developed to produce acetic acid from carbon dioxide using "cyborg bacteria".

In energy terms, natural photosynthesis can be divided in three steps:

- Light-harvesting complexes in bacteria and plants capture photons and transduce them into electrons, injecting them into the photosynthetic chain.

- Proton-coupled electron transfer along several cofactors of the photosynthetic chain, causing local, spatial charge separation.

- Redox catalysis, which uses the aforementioned transferred electrons to oxidize water to dioxygen and protons; these protons can in some species be utilized for dihydrogen production.

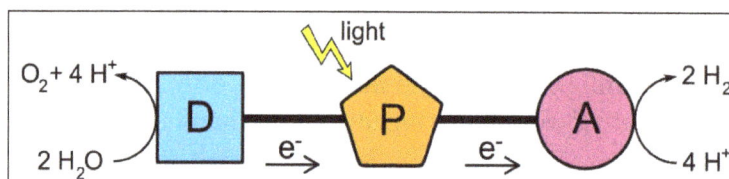

A triad assembly, with a photosensitizer (P) linked in tandem to a water oxidation catalyst (D) and a hydrogen evolving catalyst (A). Electrons flow from D to A when catalysis occurs.

Using biomimetic approaches, artificial photosynthesis tries to construct systems doing the same type of processes. Ideally, a triad assembly could oxidize water with one catalyst, reduce protons with another and have a photosensitizer molecule to power the whole system. One of the simplest designs is where the photosensitizer is linked in tandem between a water oxidation catalyst and a hydrogen evolving catalyst:

- The photosensitizer transfers electrons to the hydrogen catalyst when hit by light, becoming oxidized in the process.

- This drives the water splitting catalyst to donate electrons to the photosensitizer. In a triad assembly, such a catalyst is often referred to as a donor. The oxidized donor is able to perform water oxidation.

The state of the triad with one catalyst oxidized on one end and the second one reduced on the other end of the triad is referred to as a charge separation, and is a driving force for further electron transfer, and consequently catalysis, to occur. The different components may be assembled in diverse ways, such as supramolecular complexes, compartmentalized cells, or linearly, covalently linked molecules.

Research into finding catalysts that can convert water, carbon dioxide, and sunlight to carbohydrates or hydrogen is a current, active field. By studying the natural oxygen-evolving complex (OEC), researchers have developed catalysts such as the "blue dimer" to mimic its function or inorganic-based materials such as Birnessite with the similar building block as the OEC. Photoelectrochemical cells that reduce carbon dioxide into carbon monoxide (CO), formic acid (HCOOH) and methanol (CH_3OH) are under development. However, these catalysts are still very inefficient.

Hydrogen Catalysts

Hydrogen is the simplest solar fuel to synthesize, since it involves only the transference of two electrons to two protons. It must, however, be done stepwise, with formation of an intermediate hydride anion:

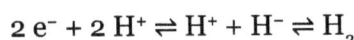

$$2\,e^- + 2\,H^+ \rightleftharpoons H^+ + H^- \rightleftharpoons H_2$$

The proton-to-hydrogen converting catalysts present in nature are hydrogenases. These are enzymes that can either reduce protons to molecular hydrogen or oxidize hydrogen to protons and electrons. Spectroscopic and crystallographic studies spanning several decades have resulted in a good understanding of both the structure and mechanism of hydrogenase catalysis. Using this information, several molecules mimicking the structure of the active site of both nickel-iron and iron-iron hydrogenases have been synthesized. Other catalysts are not structural mimics of hydrogenase but rather functional ones. Synthesized catalysts include structural H-cluster models, a dirhodium photocatalyst, and cobalt catalysts.

Water-oxidizing Catalysts

Water oxidation is a more complex chemical reaction than proton reduction. In nature, the oxygen-evolving complex performs this reaction by accumulating reducing equivalents (electrons) in a manganese-calcium cluster within photosystem II (PS II), then delivering them to water molecules, with the resulting production of molecular oxygen and protons:

$$2\,H_2O \rightarrow O_2 + 4\,H^+ + 4e^-$$

Without a catalyst (natural or artificial), this reaction is very endothermic, requiring high temperatures (at least 2500 K).

The exact structure of the oxygen-evolving complex has been hard to determine experimentally. As of 2011, the most detailed model was from a 1.9 Å resolution crystal structure of photosystem II. The complex is a cluster containing four manganese and one calcium ions, but the exact location and mechanism of water oxidation within the cluster is unknown. Nevertheless, bio-inspired manganese and manganese-calcium complexes have been synthesized, such as [Mn_4O_4] cubane-type clusters, some with catalytic activity.

Some ruthenium complexes, such as the dinuclear μ-oxo-bridged "blue dimer" (the first of its kind to be synthesized), are capable of light-driven water oxidation, thanks to being able to form high valence states. In this case, the ruthenium complex acts as both photosensitizer and catalyst.

Many metal oxides have been found to have water oxidation catalytic activity, including ruthenium(IV) oxide (RuO_2), iridium(IV) oxide (IrO_2), cobalt oxides (including nickel-doped Co_3O_4), manganese oxide (including layered MnO_2 (birnessite), Mn_2O_3), and a mix of Mn_2O_3 with $CaMn_2O_4$. Oxides are easier to obtain than molecular catalysts, especially those from relatively abundant transition metals (cobalt and manganese), but suffer from low turnover frequency and slow electron transfer properties, and their mechanism of action is hard to decipher and, therefore, to adjust.

Recently Metal-Organic Framework based materials have been shown to be a highly promising candidate for water oxidation with first row transition metals. The stability and tunability of this system is projected to be highly beneficial for future development.

Photosensitizers

Nature uses pigments, mainly chlorophylls, to absorb a broad part of the visible spectrum. Artificial systems can use either one type of pigment with a broad absorption range or combine several pigments for the same purpose.

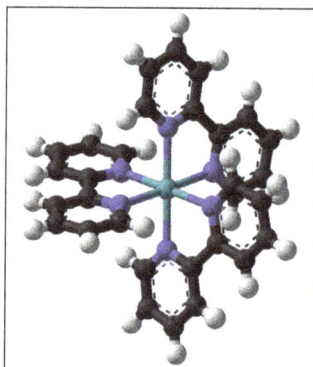

Structure of $[Ru(bipy)_3]^{2+}$, a broadly used photosensitizer.

Ruthenium polypyridine complexes, in particular tris(bipyridine)ruthenium(II) and its derivatives, have been extensively used in hydrogen photoproduction due to their efficient visible light absorption and long-lived consequent metal-to-ligand charge transfer excited state, which makes the complexes strong reducing agents. Other noble metal-containing complexes used include ones with platinum, rhodium and iridium.

Metal-free organic complexes have also been successfully employed as photosensitizers. Examples include eosin Y and rose bengal. Pyrrole rings such as porphyrins have also been used in coating nanomaterials or semiconductors for both homogeneous and heterogeneous catalysis.

As part of current research efforts artificial photonic antenna systems are being studied to determine efficient and sustainable ways to collect light for artificial photosynthesis. Gion Calzaferri describes one such antenna that uses zeolite L as a host for organic dyes, to mimic plant's light collecting systems. The antenna is fabricated by inserting dye molecules into the channels of zeolite L. The insertion process, which takes place under vacuum and at high temperature conditions, is made possible by the cooperative vibrational motion of the zeolite framework and of the dye molecules. The resulting material may be interfaced to an external device via a stopcock intermediate.

Carbon Dioxide Reduction Catalysts

In nature, carbon fixation is done by green plants using the enzyme RuBisCO as a part of the

Calvin cycle. RuBisCO is a rather slow catalyst compared to the vast majority of other enzymes, incorporating only a few molecules of carbon dioxide into ribulose-1,5-bisphosphate per minute, but does so at atmospheric pressure and in mild, biological conditions. The resulting product is further reduced and eventually used in the synthesis of glucose, which in turn is a precursor to more complex carbohydrates, such as cellulose and starch. The process consumes energy in the form of ATP and NADPH.

Artificial CO_2 reduction for fuel production aims mostly at producing reduced carbon compounds from atmospheric CO_2. Some transition metal polyphosphine complexes have been developed for this end; however, they usually require previous concentration of CO_2 before use, and carriers (molecules that would fixate CO_2) that are both stable in aerobic conditions and able to concentrate CO_2 at atmospheric concentrations haven't been yet developed. The simplest product from CO_2 reduction is carbon monoxide (CO), but for fuel development, further reduction is needed, and a key step also needing further development is the transfer of hydride anions to CO.

Other Materials and Components

Charge separation is a major property of dyad and triad assemblies. Some nanomaterials employed are fullerenes (such as carbon nanotubes), a strategy that explores the pi-bonding properties of these materials. Diverse modifications (covalent and non-covalent) of carbon nanotubes have been attempted to increase the efficiency of charge separation, including the addition of ferrocene and pyrrole-like molecules such as porphyrins and phthalocyanines.

Since photodamage is usually a consequence in many of the tested systems after a period of exposure to light, bio-inspired photoprotectants have been tested, such as carotenoids (which are used in photosynthesis as natural protectants).

Light-driven Methodologies under Development

Photoelectrochemical Cells

Photoelectrochemical cells are a heterogeneous system that use light to produce either electricity or hydrogen. The vast majority of photoelectrochemical cells use semiconductors as catalysts. There have been attempts to use synthetic manganese complex-impregnated Nafion as a working electrode, but it has been since shown that the catalytically active species is actually the broken-down complex.

A promising, emerging type of solar cell is the dye-sensitized solar cell. This type of cell still depends on a semiconductor (such as TiO_2) for current conduction on one electrode, but with a coating of an organic or inorganic dye that acts as a photosensitizer; the counter electrode is a platinum catalyst for H_2 production. These cells have a self-repair mechanism and solar-to-electricity conversion efficiencies rivaling those of solid-state semiconductor ones.

Photocatalytic Water Splitting in Homogeneous Systems

Direct water oxidation by photocatalysts is a more efficient usage of solar energy than photoelectrochemical water splitting because it avoids an intermediate thermal or electrical energy conversion step.

Bio-inspired manganese clusters have been shown to possess water oxidation activity when adsorbed on clays together with ruthenium photosensitizers, although with low turnover numbers.

As mentioned above, some ruthenium complexes are able to oxidize water under solar light irradiation. Although their photostability is still an issue, many can be reactivated by a simple adjustment of the conditions in which they work. Improvement of catalyst stability has been tried resorting to polyoxometalates, in particular ruthenium-based ones. Another way to achieve improved stability may be the use of robust clathrochelate ligands that stabilize high oxidation states of metal in catalytic intremediates.

Whereas a fully functional artificial system is usually intended when constructing a water splitting device, some mixed methods have been tried. One of these involve the use of a gold electrode to which photosystem II is linked; an electric current is detected upon illumination.

Hydrogen-producing Artificial Systems

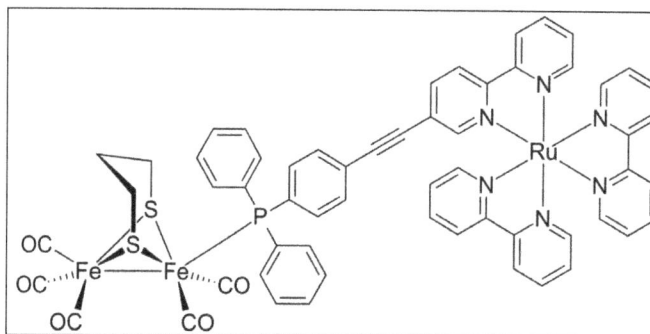

A H-cluster FeFe hydrogenase model compound covalently linked
to a ruthenium photosensitizer. The ruthenium complex absorbs light and
transduces its energy to the iron compound, which can then reduce protons to H_2.

The simplest photocatalytic hydrogen production unit consists of a hydrogen-evolving catalyst linked to a photosensitizer. In this dyad assembly, a so-called sacrificial donor for the photosensitizer is needed, that is, one that is externally supplied and replenished; the photosensitizer donates the necessary reducing equivalents to the hydrogen-evolving catalyst, which uses protons from a solution where it is immersed or dissolved in. Cobalt compounds such as cobaloximes are some of the best hydrogen catalysts, having been coupled to both metal-containing and metal-free photosensitizers. The first H-cluster models linked to photosensitizers (mostly ruthenium photosensitizers, but also porphyrin-derived ones) were prepared during the early 2000s. Both types of assembly are under development to improve their stability and increase their turnover numbers, both necessary for constructing a sturdy, long-lived solar fuel cell.

As with water oxidation catalysis, not only fully artificial systems have been idealized: hydrogenase enzymes themselves have been engineered for photoproduction of hydrogen, by coupling the enzyme to an artificial photosensitizer, such as $[Ru(bipy)_3]^{2+}$ or even photosystem I.

NADP+/NADPH Coenzyme-inspired Catalyst

In natural photosynthesis, the NADP+ coenzyme is reducible to NADPH through binding of a proton and two electrons. This reduced form can then deliver the proton and electrons, potentially as a hydride, to reactions that culminate in the production of carbohydrates (the Calvin cycle). The coenzyme is recyclable in a natural photosynthetic cycle, but this process is yet to be artificially replicated.

A current goal is to obtain an NADPH-inspired catalyst capable of recreating the natural cyclic process. Utilizing light, hydride donors would be regenerated and produced where the molecules are continuously used in a closed cycle. Brookhaven chemists are now using a ruthenium-based complex to serve as the acting model. The complex is proven to perform correspondingly with NADP+/NADPH, behaving as the foundation for the proton and two electrons needed to convert acetone to isopropanol.

Currently, Brookhaven researchers are aiming to find ways for light to generate the hydride donors. The general idea is to use this process to produce fuels from carbon dioxide.

Photobiological Production of Fuels

Some photoautotrophic microorganisms can, under certain conditions, produce hydrogen. Nitrogen-fixing microorganisms, such as filamentous cyanobacteria, possess the enzyme nitrogenase, responsible for conversion of atmospheric N_2 into ammonia; molecular hydrogen is a byproduct of this reaction, and is many times not released by the microorganism, but rather taken up by a hydrogen-oxidizing (uptake) hydrogenase. One way of forcing these organisms to produce hydrogen is then to annihilate uptake hydrogenase activity. This has been done on a strain of *Nostoc punctiforme*: one of the structural genes of the NiFe uptake hydrogenase was inactivated by insertional mutagenesis, and the mutant strain showed hydrogen evolution under illumination.

Many of these photoautotrophs also have bidirectional hydrogenases, which can produce hydrogen under certain conditions. However, other energy-demanding metabolic pathways can compete with the necessary electrons for proton reduction, decreasing the efficiency of the overall process; also, these hydrogenases are very sensitive to oxygen.

Several carbon-based biofuels have also been produced using cyanobacteria, such as 1-butanol.

Synthetic biology techniques are predicted to be useful for this topic. Microbiological and enzymatic engineering have the potential of improving enzyme efficiency and robustness, as well as constructing new biofuel-producing metabolic pathways in photoautotrophs that previously lack them, or improving on the existing ones. Another topic being developed is the optimization of photobioreactors for commercial application.

Advantages, Disadvantages and Efficiency

Advantages of solar fuel production through artificial photosynthesis include:

- The solar energy can be immediately converted and stored. In photovoltaic cells, sunlight is converted into electricity and then converted again into chemical energy for storage, with some necessary loss of energy associated with the second conversion.

- The byproducts of these reactions are environmentally friendly. Artificially photosynthesized fuel would be a carbon-neutral source of energy, which could be used for transportation or homes.

Disadvantages include:

- Materials used for artificial photosynthesis often corrode in water, so they may be less stable than photovoltaics over long periods of time. Most hydrogen catalysts are very sensitive to oxygen, being inactivated or degraded in its presence; also, photodamage may occur over time.

- The cost is not (yet) advantageous enough to compete with fossil fuels as a commercially viable source of energy.

A concern usually addressed in catalyst design is efficiency, in particular how much of the incident light can be used in a system in practice. This is comparable with photosynthetic efficiency, where light-to-chemical-energy conversion is measured. Photosynthetic organisms are able to collect about 50% of incident solar radiation, however the theoretical limit of photosynthetic efficiency is 4.6 and 6.0% for C3 and C4 plants respectively. In reality, the efficiency of photosynthesis is much lower and is usually below 1%, with some exceptions such as sugarcane in tropical climate. In contrast, the highest reported efficiency for artificial photosynthesis lab prototypes is 22.4%. However, plants are efficient in using CO_2 at atmospheric concentrations, something that artificial catalysts still cannot perform.

Importance of Photosynthesis

Photosynthesis is a crucial energy-converting process by which plants produce molecular oxygen and carbohydrates by the use of photons present in the light. The natural source of light, the sun, helps the green colored plants to fix the atmospheric carbon dioxide in to usable molecular oxygen, that we humans happen to breathe.

They help in maintaining a balanced level of oxygen and carbon dioxide in the atmosphere. Almost all the oxygen present in the atmosphere can be attributed to the process of photosynthesis, which also means that respiration and photosynthesis go together. Also, the chemical energy stored in plants is transferred to animal and humans when they consume plant matter. Photosynthesis can therefore be considered the ultimate source of life for nearly all plants and animals by providing the source of energy that drives all their metabolic processes.

References

- Listorti, andrea; durrant, james; barber, jim (december 2009). "solar to fuel". Nature materials. 8 (12): 929–930. Bibcode:2009natma...8..929l. Doi:10.1038/nmat2578. Pmid 19935695

- Photosynthesis, science: britannica.com

- Jobs (2012). "'artificial leaf' faces economic hurdle : nature news & comment". Nature. Doi:10.1038/nature.2012.10703. Retrieved 7 november 2012

- Styring, stenbjörn (21 december 2011). "artificial photosynthesis for solar fuels". Faraday discussions. 155 (advance article): 357–376. Bibcode:2012fadi..155..357s. Doi:10.1039/c1fd00113b

- man-made photosynthesis looking to change the world". Digitalworldtokyo.com. 14 january 2009. Retrieved 19 april 2011

- Importance-of-photosynthesis: whatisphotosynthesis.net, Retrieved 11 January, 2019

4

Ultraviolet Radiation

The electromagnetic radiation with a wavelength shorter than that of visible light but longer than the soft X-rays is known as ultraviolet radiation. Some of the other aspects of ultraviolet radiation are UV curing, ultraviolet germicidal irradiation, UV degradation and UV pinning. This chapter closely examines these aspects related to ultraviolet radiation to provide an extensive understanding of the subject.

Ultraviolet (UV) light is electromagnetic radiation with a wavelength shorter than that of visible light, but longer than soft X-rays. The name means "beyond violet"—*ultra* is the Latin word for "beyond", and violet is the color of the shortest wavelengths of visible light.

A collection of mineral samples brilliantly fluorescing at various wavelengths as seen while being irradiated by UV light.

Some UV wavelengths are colloquially called black light, as it is invisible to the human eye. Some animals, including birds, reptiles, and insects such as bees, can see into a portion of the ultraviolet region (the "near ultraviolet" region). Many fruits, flowers, and seeds stand out more strongly from the background in ultraviolet wavelengths as compared to human color vision. Scorpions glow or take on a yellow to green color under UV illumination. Many birds have patterns in their plumage that are invisible at usual wavelengths but observable in ultraviolet, and the urine of some animals is much easier to spot with ultraviolet.

The discovery of UV radiation was intimately associated with the observation that silver salts darken when exposed to sunlight. In 1801 German physicist Johann Wilhelm Ritter made the hallmark observation that invisible rays just beyond the violet end of the visible spectrum were especially effective at darkening paper soaked in silver chloride. He called them "deoxidizing rays" to indicate their chemical reactivity and to distinguish them from "heat rays" at the other end of the visible spectrum. The simpler term "chemical rays" was adopted shortly thereafter, and it remained popular throughout the nineteenth century. The terms chemical rays and heat rays were eventually dropped in favor of ultraviolet and infrared radiation, respectively.

Subdivisions of Ultraviolet Wavelengths

Ultraviolet radiation may be subdivided into three regions: near UV (NUV; wavelength range of 380–200 nanometers), far or vacuum UV (FUV or VUV; 200–10 nm), and extreme UV (EUV or XUV; 1–31 nm).

When considering the effect of UV radiation on human health and the environment, the range of UV wavelengths is often subdivided into UVA (400–315 nm), also called Long Wave or "blacklight"; UVB (315–280 nm), also called Medium Wave; and UVC (below 280 nm), also called Short Wave or "germicidal". In photolithography, laser technology, and similar technologies, the term deep ultraviolet (DUV) refers to wavelengths below 300 nm.

Natural Sources of UV

The Sun emits ultraviolet radiation in the UVA, UVB, and UVC bands, but because of absorption in the atmosphere's ozone layer, 99 percent of the ultraviolet radiation that reaches the Earth's surface is UVA. Some of the UVC light is responsible for the generation of the ozone.

Ordinary glass is partially transparent to UVA but is opaque to shorter wavelengths, while Silica or quartz glass, depending on quality, can be transparent even to vacuum UV wavelengths. Ordinary window glass passes about 90 percent of the light above 350 nm (nanometer, one billionth of a meter), but blocks over 90 percent of the light below 300 nm.

The onset of vacuum UV, 200 nanometers, is defined by the fact that ordinary air is opaque below this wavelength. This opacity is due to the strong absorption of light of these wavelengths by oxygen in the air. Pure nitrogen (less than about 10 parts per million oxygen) is transparent to wavelengths in the range of about 150–200 nanometers. This has wide practical significance now that semiconductor manufacturing processes are using wavelengths shorter than 200 nanometers. By working in oxygen-free gas, the equipment does not have to be built to withstand the pressure differences required to work in a vacuum.

Extreme UV (EUV) is characterized by a transition in the physics of interaction with matter: wavelengths longer than about 30 nanometers interact mainly with the valence electrons (electrons on the outermost shell of an atom) of matter, while wavelengths shorter than that interact mainly with inner shell electrons and nuclei. EUV is strongly absorbed by most known materials, but it is possible to synthesize multilayer optics that reflect up to about 50 percent of XUV radiation at normal incidence. This technology has been used to make telescopes for solar imaging; it was pioneered by the Normal Incidence X-Ray Telescope (NIXT) and Multi Spectral Telescope Array (MSSTA) sounding rockets in the 1990s; (current examples are SOHO/EIT and TRACE) and for nanolithography (printing of traces and devices on microchips).

Beneficial Effects

A positive effect of UVB light is that it induces the production of vitamin D in the skin. It has been estimated that tens of thousands of premature deaths occur in the US annually from a range of cancers due to insufficient UVB exposure (via vitamin D deficiency). Another effect of vitamin D deficiency is osteomalacia (rickets), which can result in bone pain, difficulty in weight bearing and sometimes fractures.

Ultraviolet radiation has other medical applications, in the treatment of skin conditions such as psoriasis and vitiligo. UVB and UVA radiation can be used, in conjunction with psoralens (PUVA) treatment. In cases of psoriasis and vitiligo UV light with wavelength of 311 nanometers is most effective.

Safety Aspects of UV

In humans, prolonged exposure to solar UV radiation may result in acute and chronic health effects on the skin, eye, and immune system. UVC rays are the highest energy, most dangerous type of ultraviolet light. Little attention has been given to UVC rays in the past since they are filtered out by the atmosphere. However, their use in equipment such as pond sterilization units may pose an exposure risk, if the lamp is switched on outside of its enclosed pond sterilization unit.

Skin

"Ultraviolet (UV) irradiation present in sunlight is an environmental human carcinogen. The toxic effects of UV from natural sunlight and therapeutic artificial lamps are a major concern for human health. The major acute effects of UV irradiation on normal human skin comprise sunburn inflammation (erythema), tanning, and local or systemic immunosuppression."

— Y. Matsumura and H. N. Ananthaswamy

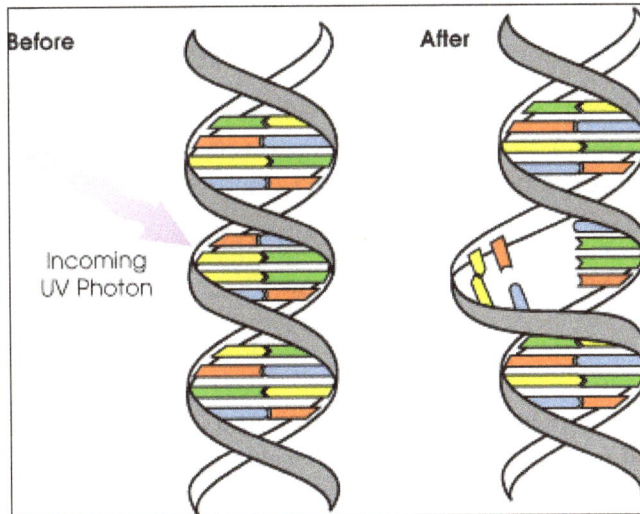

Ultraviolet photons harm the DNA molecules of living organisms in different ways. In one common damage event, adjacent bases bond with each other, instead of across the "ladder". This makes a bulge, and the distorted DNA molecule does not function properly.

Prolonged exposure to UVA, UVB and UVC can all damage collagen fibers and thereby accelerate aging of the skin. In general, UVA is the least harmful, but can contribute to the aging of skin, DNA damage and possibly skin cancer. It penetrates deeply and does not cause sunburn. Because it does not cause reddening of the skin (erythema) it cannot be measured in the sun protection factor testing. There is no good clinical measurement of the blocking of UVA radiation, but it is important that sunscreen block both UVA and UVB.

UVA light is also known as "black light" and, because of its longer wavelength, can penetrate many windows. It also penetrates deeper into the skin than UVB light and is thought to be a prime cause of wrinkles.

UVB light can cause skin cancer (if there is prolonged exposure). The radiation excites DNA molecules in skin cells, causing covalent bonds to form between adjacent thymine bases, producing thymidine dimers. Thymidine dimers do not base pair normally, which can cause distortion of the DNA helix, stalled replication, gaps, and misincorporation. These can lead to mutations, which can result in cancerous growths. The mutagenicity of UV radiation can be easily observed in bacteria cultures. This cancer connection is one reason for concern about ozone depletion and the ozone hole.

As a defense against UV radiation, the body tans when exposed to moderate (depending on skin type) levels of radiation by releasing the brown pigment melanin. This helps to block UV penetration and prevent damage to the vulnerable skin tissues deeper down. Suntan lotion that partly blocks UV is widely available (often referred to as "sun block" or "sunscreen"). Most of these products contain an "SPF rating" that describes the amount of protection given. This protection, however, applies only to UVB rays responsible for sunburn and not to UVA rays that penetrate more deeply into the skin and may also be responsible for causing cancer and wrinkles. Some sunscreen lotion now includes compounds such as titanium dioxide which helps protect against UVA rays. Other UVA blocking compounds found in sunscreen include zinc oxide and avobenzone. There are also naturally occurring compounds found in rainforest plants that have been known to protect the skin from UV radiation damage, such as the fern *Phlebodium aureum*.

What to look for in sunscreen: UVB protection: Padimate O, Homosalate, Octisalate (octyl salicylate), Octinoxate (octyl methoxycinnamate) UVA protection: Avobenzone UVA/UVB protection: Octocrylene, titanium dioxide, zinc oxide, Mexoryl (ecamsule).

Another means to block UV is sun protective clothing. This is clothing that has a "UPF rating" that describes the protection given against both UVA and UVB.

Eye

High intensities of UVB light are hazardous to the eyes, and exposure can cause *welder's flash* (photokeratitis or arc eye) and may lead to cataracts, pterygium, and pinguecula formation.

Protective eyewear is beneficial to those who are working with or those who might be exposed to ultraviolet radiation, particularly short wave UV. Given that light may reach the eye from the sides, full coverage eye protection is usually warranted if there is an increased risk of exposure, as in high altitude mountaineering. Mountaineers are exposed to higher than ordinary levels of UV radiation, both because there is less atmospheric filtering and because of reflection from snow and ice.

Ordinary, untreated eyeglasses give some protection. Most plastic lenses give more protection than glass lenses, because, as noted above, glass is transparent to UVA and the common acrylic plastic used for lenses is less so. Some plastic lens materials, such as polycarbonate, inherently block most UV. There are protective treatments available for eyeglass lenses that need it which will give better protection. But even a treatment that *completely* blocks UV will not protect the eye from light that arrives around the lens. To convince yourself of the potential dangers of stray UV light, cover your lenses with something opaque, like aluminum foil, stand next to a bright light, and consider how much light you see, despite the complete blockage of the lenses. Most intraocular lenses help to protect the retina by absorbing UV radiation.

Applications of UV

Black Lights

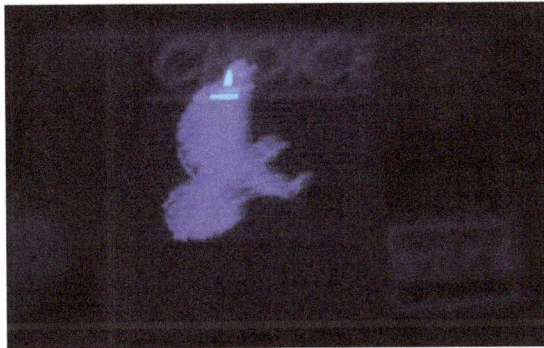

A bird appears on every Visa credit card when held under a UV light source.

A black light is a lamp that emits long wave UV radiation and very little visible light. Fluorescent black lights are typically made in the same fashion as normal fluorescent lights except that only one phosphor is used and the normally clear glass envelope of the bulb is replaced by a deep bluish purple glass called Wood's glass.

To thwart counterfeiters, sensitive documents (e.g. credit cards, driver's licenses, passports) may also include a UV watermark that can only be seen when viewed under a UV-emitting light. Passports issued by most countries usually contain UV sensitive inks and security threads. Visa stamps and stickers such as those issued by Ukraine contain large and detailed seals invisible to the naked eye under normal lights, but strongly visible under UV illimunation. Passports issued by the United States have the UV sensitive threads on the last page of the passport along with the barcode.

Fluorescent Lamps

Fluorescent lamps produce UV radiation by ionising low-pressure mercury vapour. A phosphorescent coating on the inside of the tubes absorbs the UV and converts it to visible light. The main mercury emission wavelength is in the UVC range. Unshielded exposure of the skin or eyes to mercury arc lamps that do not have a conversion phosphor is quite dangerous. Other practical UV sources with more continuous emission spectra include xenon arc lamps (commonly used as sunlight simulators), deuterium arc lamps, mercury-xenon arc lamps, metal-halide arc lamps, and tungsten-halogen incandescent lamps.

Astronomy

Aurora at Jupiter's north pole as seen in ultraviolet light by the Hubble Space Telescope.

In astronomy, very hot objects preferentially emit UV radiation. However, the same ozone layer that protects us causes difficulties for astronomers observing from the Earth, so most UV observations are made from space.

Pest Control

Ultraviolet fly traps are used for the elimination of various small flying insects. They are attracted to the UV light and are killed using an electrical shock or trapped once they come into contact with the device.

Spectrophotometry

UV/VIS spectroscopy is widely used as a technique in chemistry, for analysis of chemical structure, most notably conjugated systems (a system of atoms with alternating single and double covalent bonds). UV radiation is often used in visible spectrophotometry to determine the existence of fluorescence in a given sample.

Analyzing Minerals

Ultraviolet lamps are also used in analyzing minerals, gems, and in other detective work including authentication of various collectibles. Materials may look the same under visible light, but fluoresce to different degrees under ultraviolet light; or may fluoresce differently under short wave ultraviolet versus long wave ultraviolet. UV fluorescent dyes are used in many applications (for example, biochemistry and forensics). The fluorescent protein Green Fluorescent Protein (GFP) is often used in genetics as a marker. Many substances, proteins for instance, have significant light absorption bands in the ultraviolet that are of use and interest in biochemistry and related fields. UV-capable spectrophotometers are common in such laboratories.

Photolithography

Ultraviolet radiation is used for very fine resolution photolithography, a procedure where a chemical known as a photoresist is exposed to UV radiation which has passed through a mask. The light allows chemical reactions to take place in the photoresist, and after development (a step that either removes the exposed or unexposed photoresist), a geometric pattern which is determined by the mask remains on the sample. Further steps may then be taken to "etch" away parts of the sample with no photoresist remaining.

UV radiation is used extensively in the electronics industry because photolithography is used in the manufacture of semiconductors, integrated circuit components, and printed circuit boards.

Checking Electrical Insulation

A new application of UV is to detect corona discharge (often simply called "corona") on electrical apparatus. Degradation of insulation of electrical apparatus or pollution causes corona, wherein a strong electric field ionizes the air and excites nitrogen molecules, causing the emission of ultraviolet radiation. Corona produces ozone and to a lesser extent nitrogen oxide which may subsequently react with water in the air to form nitrous acid and nitric acid vapour in the surrounding air.

Sterilization

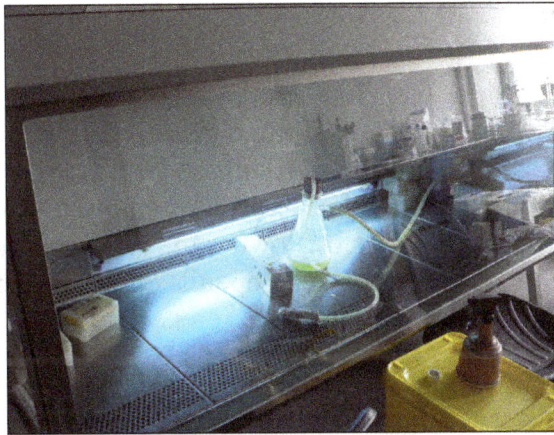

A low pressure mercury vapor discharge tube floods the inside of a hood with shortwave UV light when not in use, sterilizing microbiological contaminants from irradiated surfaces.

Ultraviolet lamps are used to sterilize workspaces and tools used in biology laboratories and medical facilities. Commercially-available low pressure mercury-vapor lamps emit about 86 percent of their light at 254 nanometers, which coincides very well with one of the two peaks of the germicidal effectiveness curve (i.e., effectiveness of UV absorption by DNA). One of these peaks is at about 265 nanometers and the other is at about 185 nanometers. Although 185 nanometers is better absorbed by DNA, the quartz glass used in commercially-available lamps, as well as environmental media such as water, are more opaque to 185 nanometers than 254 nanometers. UV light at these germicidal wavelengths causes adjacent thymine molecules on DNA to dimerize, if enough of these defects accumulate on a microorganism's DNA its replication is inhibited, thereby rendering it harmless (even though the organism may not be killed outright). Since microorganisms can be shielded from ultraviolet light in small cracks and other shaded areas, however, these lamps are used only as a supplement to other sterilization techniques.

Disinfecting Drinking Water

UV radiation can be an effective viricide and bactericide. Disinfection using UV radiation was more commonly used in wastewater treatment applications but is finding increased usage in drinking water treatment. A process named SODIS has been extensively researched in Switzerland and proven ideal to treat small quantities of water. Contaminated water is filled into transparent plastic bottles and exposed to full sunlight for six hours. The sunlight is treating the contaminated water through two synergetic mechanisms: Radiation in the spectrum of UV-A (wavelength 320-400 nanometers) and increased water temperature. If the water temperatures raises above 50 °C, the disinfection process is three times faster.

It used to be thought that UV disinfection was more effective for bacteria and viruses, which have more exposed genetic material, than for larger pathogens which have outer coatings or that form cyst states (e.g., Giardia) that shield their DNA from the UV light. However, it was recently discovered that ultraviolet radiation can be somewhat effective for treating the microorganism cryptosporidium. The findings resulted in two U.S. patents and the use of UV radiation as a viable method to treat drinking water. Giardia, in turn, has been shown to be very susceptible to UVC when the tests were based on infectivity rather than excystation. It turns out that protists are able to survive high UVC doses but are sterilized at low doses.

Food Processing

As consumer demand for fresh and "fresh like" food products increases, the demand for nonthermal methods of food processing is likewise on the rise. In addition, public awareness regarding the dangers of food poisoning is also raising the demand for improved food processing methods. Ultraviolet radiation is used in several food processes to remove unwanted microorganisms. UV light can be used to pasteurize fruit juices by flowing the juice over a high intensity ultraviolet light source. The effectiveness of such a process depends on the UV absorbance of the juice.

Fire Detection

Ultraviolet detectors generally use either a solid-state device, such as one based on silicon carbide or aluminum nitride, or a gas-filled tube as the sensing element. UV detectors which are sensitive to UV light in any part of the spectrum respond to irradiation by sunlight and artificial light. A burning hydrogen flame, for instance, radiates strongly in the 185 to 260 nanometer range and only very weakly in the infrared (IR) region, while a coal fire emits very weakly in the UV band yet very strongly at IR wavelengths; thus a fire detector which operates using both UV and IR detectors is more reliable than one with a UV detector alone. Virtually all fires emit some radiation in the UVB band, while the Sun's radiation at this band is absorbed by the Earth's atmosphere. The result is that the UV detector is "solar blind," meaning it will not cause an alarm in response to radiation from the Sun, so it can easily be used both indoors and outdoors.

UV detectors are sensitive to most fires, including hydrocarbons, metals, sulfur, hydrogen, hydrazine, and ammonia. Arc welding, electrical arcs, lightning, X-rays used in nondestructive metal testing equipment (though this is highly unlikely), and radioactive materials can produce levels that will activate a UV detection system. The presence of UV-absorbing gases and vapors will attenuate the UV radiation from a fire, adversely affecting the ability of the detector to detect flames. Likewise, the presence of an oil mist in the air or an oil film on the detector window will have the same effect.

Curing of Inks, Adhesives and Coatings

Certain inks, coatings, and adhesives are formulated with photoinitiators (ingredient that absorbs light) and resins. When exposed to the correct energy and irradiance in the required band of UV light, polymerisation occurs, and so the adhesives harden or cure. Usually, this reaction is very quick, a matter of a few seconds. Applications include glass and plastic bonding, optical fiber coatings, the coating of flooring, paper finishes in offset printing, and dental fillings.

An industry has developed around the manufacture of UV sources for UV curing applictions. Fast processes such as flexo or offset printing require high intensity light focused via reflectors onto a moving substrate and medium and high pressure mercury- or iron-based bulbs can be energised with electric arc or microwaves. Lower power fluorescent lamps can be used for static applications and in some cases, small high pressure lamps can have light focussed and transmitted to the work area via liquid filled or fibre optic light guides.

Deterring Substance Abuse in Public Places

UV lights have been installed in some parts of the world in public restroom and on public transport for the purpose of deterring substance abuse. The blue color of these lights, combined with the fluorescence of the skin, make it harder for intravenous drug users to find a vein. The efficacy of these lights for that purpose has been questioned, with some suggesting that drug users simply find a vein outside the public restroom and mark the spot with a marker for accessibility when inside the restroom. There is currently no published evidence supporting the idea of a deterrent effect.

Erasing EPROM Modules

Some EPROM (electronically programmable read-only memory) modules are erased by exposure to UV radiation. These modules often have a transparent glass (quartz) window on the top of the chip that allows the UV radiation in. These have been largely superseded by EEPROM and flash memory chips in most devices.

Preparing Low Surface Energy Polymers

UV radiation is useful in preparing low surface energy polymers for adhesives. Polymers exposed to UV light will oxidize thus raising the surface energy of the polymer. Once the surface energy of the polymer has been raised, the bond between the adhesive and the polymer will be greater.

Reading Completely Illegible Papyruses

Using multi-spectral imaging it is possible to read illegible papyruses, such as the burned papyruses of the Villa of the Papyri or of Oxyrhynchus. The technique involves taking pictures of the illegible papyruses using different filters in the infrared or ultraviolet range, finely tuned to capture certain wavelengths of light. Thus, the optimum spectral portion can be found for distinguishing ink from paper on the papyrus surface.

UV Curing

UV curing is the process by which ultraviolet light is used to initiate a photochemical reaction that generates a crosslinked network of polymers. UV curing is adaptable to printing, coating, decorating, stereolithography, and in the assembly of a variety of products and materials. In comparison to other technologies, curing with UV energy may be considered a low temperature process, a high speed process, and is a solventless process, as cure occurs via direct polymerization rather than by evaporation. Originally introduced in the 1960s, this technology has streamlined and increased automation in many industries in the manufacturing sector.

Applications

UV curing is used whenever there is a need for curing and drying of inks, adhesives and coatings. UV-cured adhesive has become a high-speed replacement for two-part adhesives, eliminating the need for solvent removal, ratio mixing and potential life concern. It is used in the screen printing process, where UV curing systems are used to polymerize images on screen-printed products, ranging from T-shirts to 3D and cylindrical parts. It is used in fine instrument finishing (guitars, violins, ukuleles, etc.), pool cue manufacturing and other wood craft industries. Printing with UV curable inks provides the ability to print on a very wide variety of substrates such as plastics, paper, canvas, glass, metal, foam boards, tile, films, and many other materials.

Other industries that take advantage of UV curing include medicine, automobiles, cosmetics (for example artificial fingernails and gel nail polish), food, science, education and art.

Advantages of UV Curing

The primary advantage of curing finishes and inks with ultraviolet light is the speed at which the final product can be readied for shipping. In addition to speeding up production, this can also reduce flaws and errors as the amount of time that dust, flies or any airborne object has to settle upon the object is reduced. This can increase the quality of the finished item, and allow for greater consistency.

The other obvious benefit is that manufacturers can devote less space to finishing items, since they don't have to wait for them to dry. This creates an efficiency that ripples through the entire manufacturing process.

Types of UV Curing

Mercury vapor lamps are the industry standard for curing products with ultraviolet light. The bulbs work by high voltage passing through, vaporizing the mercury. An arc is created within the mercury which emits a spectral output in the UV region of the light spectrum. The light intensity occurs in the 240-270nm and 350-380nm. This intense spectrum of light is what causes the rapid curing of the different applications being used.

In the last few years an emerging type of UV curing technology called UV LED curing has entered the marketplace. This technology is growing rapidly in popularity and has many advantages over mercury based lamps although is not the right fit for every application.

Fluorescent lamps made specifically for UV curing are also available. These have the ability to dial into specific frequencies at a lower price point as fluorescent lamps are an established technology and the spectrum is easily controlled by the type of phosphor used. They can produce frequencies that LEDs and mercury vapor lamps can not, including multiple frequencies. They are somewhat less efficient than LEDs or Mercury vapor but cost a fraction of the price of the other systems. They allow for curing all around an item by using multiple tubes and off the shelf ballast systems.

Types of Ultraviolet Lamps

Mercury Vapor Lamp (H Type)

The mercury lamp has an output in the short wave UV range between 220 and 320 nm (nanometers) and a spike of energy in the longwave range at 365 nm. The H lamp is a good choice for clear coatings and thin ink layers and produces hard surface cures and high gloss finishes.

Mercury Vapor Lamp with Iron Additive (D Type)

The addition of iron to the lamp yields a strong output in the longwave range between 350 and 400 nm while the mercury component maintains good output in the short wavelength range. The D lamp is a good choice for curing heavily pigmented inks, adhesives, and thick laydowns of clear materials.

Mercury Vapor Lamp with Gallium Additive (V Type)

The addition of gallium to the lamp yields a strong output in the longwave range between 400 and 450 nm. This makes the V lamp a good choice for curing white pigmented inks and base coats containing titanium dioxide which blocks the most shortwave UV.

Fluorescent Lamps

Fluorescent lamps are used for UV curing in a number of applications. In particular, these are used where the excessive heat of mercury vapor is undesirable, or when an item needs more than a single source of light and instead the item needs to be surrounded by light, such as musical instruments. Fluorescent lamps can be created that produce ultraviolet anywhere within the UVA/UVB spectrum. Additionally, lamps that have multiple peaks are possible, allowing a wider variety of photoinitiators to be used. While fluorescent lamps are less efficient at producing UV than mercury vapor, newer initiators require less total energy, offsetting this disadvantage. Fluorescent lamps in a wide variety of sizes and wattages are available.

LEDs

UV LED devices are capable of emitting a narrow spectrum of radiation (+/- 10 nm), while mercury lamps have a broader spectral distribution. Fluorescent ultraviolet lamps can be fairly narrow, although not as narrow as LEDs.

LEDs are much more expensive but last up to 10 times longer, and unlike fluorescent tubes, can be cycled on and off frequently as they require no startup or cool down period. While they can not produce the same spectrum as mercury vapor or fluorescent tubes, photoinitiators can be formulated to work with them easily. Other advantages of UV LED curing systems are the ability to be more compact, the ability to work with heat-sensitive substrates, better energy efficiency and improved safety by avoiding use of mercury.

Ultraviolet Germicidal Irradiation

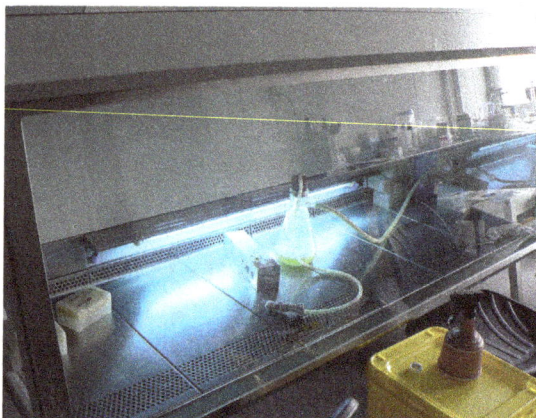

A low-pressure mercury-vapor discharge tube floods the inside of
a biosafety cabinet with shortwave UV light when not in use,
sterilizing microbiological contaminants from irradiated surfaces.

Ultraviolet germicidal irradiation (UVGI) is a disinfection method that uses short-wavelength ultraviolet (UV-C) light to kill or inactivate microorganisms by destroying nucleic acids and disrupting their DNA, leaving them unable to perform vital cellular functions. UVGI is used in a variety of applications, such as food, air, and water purification.

UV-C light is weak at the Earth's surface as the ozone layer of the atmosphere blocks it. UVGI devices can produce strong enough UV-C light in circulating air or water systems to make them inhospitable environments to microorganisms such as bacteria, viruses, molds and other pathogens. UVGI can be coupled with a filtration system to sanitize air and water.

The application of UVGI to disinfection has been an accepted practice since the mid-20th century. It has been used primarily in medical sanitation and sterile work facilities. Increasingly it has been employed to sterilize drinking and wastewater, as the holding facilities are enclosed and can be circulated to ensure a higher exposure to the UV. In recent years UVGI has found renewed application in air purifiers.

Method of Operation

UV light is electromagnetic radiation with wavelengths shorter than visible light but longer than X-rays. UV can be separated into various ranges, with short-wavelength UV (UVC) considered "germicidal UV". Wavelengths between about 200 nm and 300 nm are strongly absorbed by nucleic acids. The absorbed energy can result in defects including pyrimidine dimers. These dimers can prevent replication or can prevent the expression of necessary proteins, resulting in the death or inactivation of the organism.

- Mercury-based lamps operating at low vapor pressure emit UV light at the 253.7 nm line.

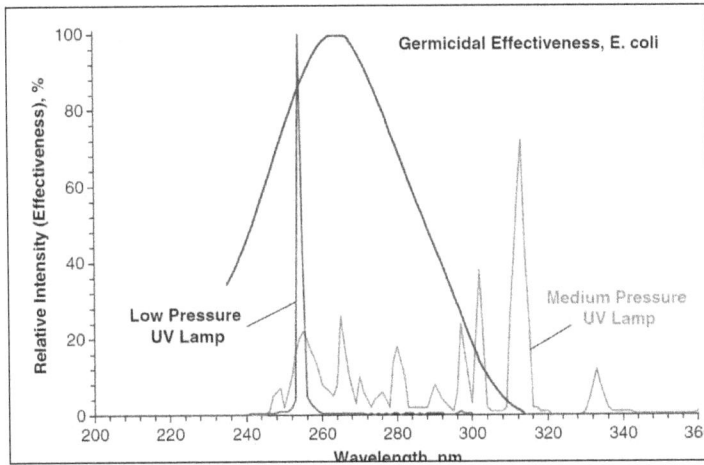

Low-pressure & medium-pressure mercury lamps compared to E.coli germicidal
effectiveness curve. Ultraviolet Germicidal Irradiation Handbook.

- Ultraviolet light-emitting diodes (UV-C LED) lamps emit UV light at selectable wavelengths between 255 and 280 nm.

- Pulsed-xenon lamps emit UV light across the entire UV spectrum with a peak emission near 230 nm.

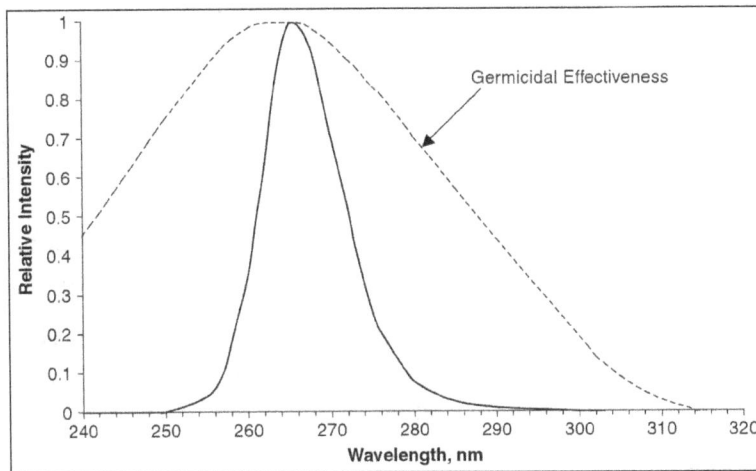

UV-C LED emitting 265 nm compared to E.coli germicidal effectiveness curve.
Ultraviolet Germicidal Irradiation Handbook.

This process is similar to the effect of longer wavelengths (UVB) producing sunburn in humans. Microorganisms have less protection against UV, and cannot survive prolonged exposure to it.

A UVGI system is designed to expose environments such as water tanks, sealed rooms and forced air systems to germicidal UV. Exposure comes from germicidal lamps that emit germicidal UV at the correct wavelength, thus irradiating the environment. The forced flow of air or water through this environment ensures exposure.

Effectiveness

The effectiveness of germicidal UV depends on the length of time a microorganism is exposed to

UV, the intensity and wavelength of the UV radiation, the presence of particles that can protect the microorganisms from UV, and a microorganism's ability to withstand UV during its exposure.

In many systems, redundancy in exposing microorganisms to UV is achieved by circulating the air or water repeatedly. This ensures multiple passes so that the UV is effective against the highest number of microorganisms and will irradiate resistant microorganisms more than once to break them down.

"Sterilization" is often misquoted as being achievable. While it is theoretically possible in a controlled environment, it is very difficult to prove and the term "disinfection" is generally used by companies offering this service as to avoid legal reprimand. Specialist companies will often advertise a certain log reduction, e.g., 6-log reduction or 99.9999% effective, instead of sterilization. This takes into consideration a phenomenon known as light and dark repair (photoreactivation and base excision repair, respectively), in which a cell can repair DNA that has been damaged by UV light.

The effectiveness of this form of disinfection depends on line-of-sight exposure of the microorganisms to the UV light. Environments where design creates obstacles that block the UV light are not as effective. In such an environment, the effectiveness is then reliant on the placement of the UVGI system so that line of sight is optimum for disinfection.

Dust and films coating the bulb lower UV output. Therefore, bulbs require periodic cleaning and replacement to ensure effectiveness. The lifetime of germicidal UV bulbs varies depending on design. Also, the material that the bulb is made of can absorb some of the germicidal rays.

Lamp cooling under airflow can also lower UV output; thus, care should be taken to shield lamps from direct airflow, or to add additional lamps to compensate for the cooling effect.

Increases in effectiveness and UV intensity can be achieved by using reflection. Aluminum has the highest reflectivity rate versus other metals and is recommended when using UV.

One method for gauging UV effectiveness in water disinfection applications is to compute UV dose. The U.S. EPA publishes UV dosage guidelines for water treatment applications. UV dose cannot be measured directly but can be inferred based on the known or estimated inputs to the process:

- Flow rate (contact time).

- Transmittance (light reaching the target).

- Turbidity (cloudiness).

- Lamp age or fouling or outages (reduction in UV intensity).

In air and surface disinfection applications the UV effectiveness is estimated by calculating the UV dose which will be delivered to the microbial population. The UV dose is calculated as follows:

$$\text{UV dose } \mu Ws/cm^2 = \text{UV intensity } \mu W/cm^2 \times \text{Exposure time (seconds)}$$

The UV intensity is specified for each lamp at a distance of 1 meter. UV intensity is inversely proportional to the square of the distance so it decreases at longer distances. Alternatively, it rapidly increases at distances shorter than 1m. In the above formula the UV intensity must always be adjusted for distance unless the UV dose is calculated at exactly 1m from the lamp. Also, to ensure effectiveness the UV dose must be calculated at the end of lamp life (EOL is specified in number of hours when the lamp is expected to reach 80% of its initial UV output) and at the furthest distance from the lamp on the periphery of the target area. Some *shatter-proof* lamps are coated with a fluorated ethylene polymer to contain glass shards and mercury in case of breakage; this coating reduces UV output by as much as 20%.

To accurately predict what UV dose will be delivered to the target the UV intensity, adjusted for distance, coating and end of lamp life, will be multiplied by the exposure time. In static applications the exposure time can be as long as needed for an effective UV dose to be reached. In case of rapidly moving air, in AC air ducts for example, the exposure time is short so the UV intensity must be increased by introducing multiple UV lamps or even banks of lamps. Also, the UV installation must be located in a long straight duct section with the lamps perpendicular to the air flow to maximize the exposure time.

These calculations actually predict the UV fluence and it is assumed that the UV fluence will be equal to the UV dose. The UV dose is the amount of germicidal UV energy absorbed by a microbial population over a period of time. If the microorganisms are planktonic (free floating) the UV fluence will be equal the UV dose. However, if the microorganisms are protected by mechanical particles, such as dust and dirt, or have formed biofilm a much higher UV fluence will be needed for an effective UV dose to be introduced to the microbial population.

Inactivation of Microorganisms

The degree of inactivation by ultraviolet radiation is directly related to the UV dose applied to the water. The dosage, a product of UV light intensity and exposure time, is usually measured in microjoules per square centimeter, or equivalently as microwatt seconds per square centimeter ($\mu W \cdot s/cm^2$). Dosages for a 90% kill of most bacteria and viruses range from 2,000 to 8,000 $\mu W \cdot s/cm^2$. Larger parasites such as cryptosporidium require a lower dose for inactivation. As a result, the U.S. Environmental Protection Agency has accepted UV disinfection as a method for drinking water plants to obtain cryptosporidium, giardia or virus inactivation credits. For example, for a 90% reduction of cryptosporidium, a minimum dose of 2,500 $\mu W \cdot s/cm^2$ is required based on the U.S. EPA UV Guidance Manual published in 2006.

Strengths and Weaknesses

Advantages

UV water treatment devices can be used for well water and surface water disinfection. UV treatment compares favorably with other water disinfection systems in terms of cost, labor, and the need for technically trained personnel for operation. Water chlorination treats larger organisms and offers residual disinfection, but these systems are expensive because they need special operator training and a steady supply of a potentially hazardous material. Finally, boiling of water is the most reliable treatment method but it demands labor, and imposes a high economic cost. UV treatment is rapid and, in terms of primary energy use, approximately 20,000 times more efficient than boiling.

Disadvantages

UV disinfection is most effective for treating high-clarity, purified reverse osmosis distilled water. Suspended particles are a problem because microorganisms buried within particles are shielded from the UV light and pass through the unit unaffected. However, UV systems can be coupled with a pre-filter to remove those larger organisms that would otherwise pass through the UV system unaffected. The pre-filter also clarifies the water to improve light transmittance and therefore UV dose throughout the entire water column. Another key factor of UV water treatment is the flow rate—if the flow is too high, water will pass through without sufficient UV exposure. If the flow is too low, heat may build up and damage the UV lamp.

A disadvantage of UVGI is that while water treated by chlorination is resistant to reinfection (until the chlorine off-gasses), UVGI water is not resistant to reinfection. UVGI water must be transported or delivered in such a way as to avoid reinfection.

Safety

In UVGI systems the lamps are shielded or are in environments that limit exposure, such as a closed water tank or closed air circulation system, often with interlocks that automatically shut off the UV lamps if the system is opened for access by humans.

For human beings, skin exposure to germicidal wavelengths of UV light can produce rapid sunburn and skin cancer. Exposure of the eyes to this UV radiation can produce extremely painful inflammation of the cornea and temporary or permanent vision impairment, up to and including blindness in some cases. UV can damage the retina of the eye.

Another potential danger is the UV production of ozone, which can be harmful to health. The US Environmental Protection Agency designated 0.05 parts per million (ppm) of ozone to be a safe level. Lamps designed to release UVC and higher frequencies are doped so that any UV light below 254 nm wavelengths will not be released, to minimize ozone production. A full-spectrum lamp will release all UV wavelengths, and will produce ozone when UVC hits oxygen (O_2) molecules.

UV-C radiation is able to break down chemical bonds. This leads to rapid aging of plastics, insulation, gaskets, and other materials. Note that plastics sold to be "UV-resistant" are tested only for UV-B, as UV-C doesn't normally reach the surface of the Earth. When UV is used near plastic, rubber, or insulation, care should be taken to shield these items; metal tape or aluminum foil will suffice.

The American Conference of Governmental Industrial Hygienists (ACGIH) Committee on Physical Agents has established a TLV for UV-C exposure to avoid such skin and eye injuries among those most susceptible. For 254 nm UV, this TLV is 6 mJ/cm² over an eight-hour period. The TLV function differs by wavelengths because of variable energy and potential for cell damage. This TLV is supported by the International Commission on Non-Ionizing Radiation Protection and is used in setting lamp safety standards by the Illuminating Engineering Society of North America. When TUSS was planned, and until quite recently, this TLV was interpreted as if eye exposure in rooms was continuous over eight hours and at the highest eye-level irradiance found in the room. In those highly unlikely conditions, a 6.0 mJ/cm² dose is reached under the ACGIH TLV after just eight hours of continuous exposure to an irradiance of 0.2 µW/cm². Thus, 0.2 µW/cm² was widely interpreted as the upper permissible limit of irradiance at eye height.

Uses

Air Disinfection

UVGI can be used to disinfect air with prolonged exposure. Disinfection is a function of UV intensity and time. For this reason, it is not as effective on moving air, or when the lamp is perpendicular to the flow, as exposure times are dramatically reduced. Air purification UVGI systems can be free-standing units with shielded UV lamps that use a fan to force air past the UV light. Other systems are installed in forced air systems so that the circulation for the premises moves microorganisms past the lamps. Key to this form of sterilization is placement of the UV lamps and a good filtration system to remove the dead microorganisms. For example, forced air systems by design impede line-of-sight, thus creating areas of the environment that will be shaded from the UV light. However, a UV lamp placed at the coils and drain pans of cooling systems will keep microorganisms from forming in these naturally damp places.

ASHRAE covers UVGI and its applications in indoor air quality and building maintenance in "Ultraviolet Lamp Systems", Chapter 16 of its 2008 Handbook, *HVAC Systems and Equipment*. Its 2011 Handbook, *HVAC Applications*, covers "Ultraviolet air and surface treatment" in Chapter 60.

Water Disinfection

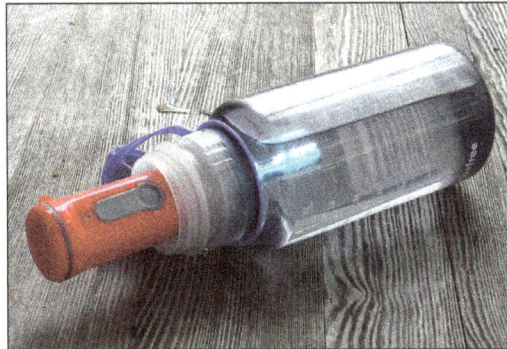

A portable, battery-powered, low-pressure
mercury-vapor discharge lamp for water sterilization.

Ultraviolet disinfection of water is a purely physical, chemical-free process. Even parasites such as *cryptosporidia* or *giardia*, which are extremely resistant to chemical disinfectants, are efficiently reduced. UV can also be used to remove chlorine and chloramine species from water; this process is called photolysis, and requires a higher dose than normal disinfection. The sterilized microorganisms are not removed from the water. UV disinfection does not remove dissolved organics, inorganic compounds or particles in the water. The world's largest water disinfection plant treats drinking water for New York city. The Catskill-Delaware Water Ultraviolet Disinfection Facility, commissioned on 8 October 2013, incorporates a total of 56 energy-efficient UV reactors treating up to 2.2 billion US gallons (8,300,000 m^3) a day.

Ultraviolet can also be combined with ozone or hydrogen peroxide to produce hydroxyl radicals to break down trace contaminants through an Advanced oxidation process.

It used to be thought that UV disinfection was more effective for bacteria and viruses, which have more-exposed genetic material, than for larger pathogens that have outer coatings or that form cyst states (e.g., Giardia) that shield their DNA from UV light. However, it was recently discovered that ultraviolet radiation can be somewhat effective for treating the microorganism Cryptosporidium. The findings resulted in the use of UV radiation as a viable method to treat drinking water. Giardia in turn has been shown to be very susceptible to UV-C when the tests were based on infectivity rather than excystation. It has been found that protists are able to survive high UV-C doses but are sterilized at low doses.

Developing Countries

A 2006 project at University of California, Berkeley produced a design for inexpensive water disinfection in resource deprived settings. The project was designed to produce an open source design that could be adapted to meet local conditions. In a somewhat similar proposal in 2014, Australian students designed a system using chip packet foil to reflect solar UV radiation into a glass tube that should disinfect water without power.

Wastewater Treatment

Ultraviolet in sewage treatment is commonly replacing chlorination. This is in large part because of concerns that reaction of the chlorine with organic compounds in the waste water stream could synthesize potentially toxic and long lasting chlorinated organics and also because of the environmental risks of storing chlorine gas or chlorine containing chemicals. Individual wastestreams to be treated by UVGI must be tested to ensure that the method will be effective due to potential interferences such as suspended solids, dyes, or other substances that may block or absorb the UV radiation. According to the World Health Organization, "UV units to treat small batches (1 to several liters) or low flows (1 to several liters per minute) of water at the community level are estimated to have costs of US$20 per megaliter, including the cost of electricity and consumables and the annualized capital cost of the unit."

Large-scale urban UV wastewater treatment is performed in cities such as Edmonton, Alberta. The use of ultraviolet light has now become standard practice in most municipal wastewater treatment processes. Effluent is now starting to be recognized as a valuable resource, not a problem that needs to be dumped. Many wastewater facilities are being renamed as water reclamation facilities, whether the wastewater is discharged into a river, used to irrigate crops, or injected into an aquifer for later recovery. Ultraviolet light is now being used to ensure water is free from harmful organisms.

Aquarium and Pond

Ultraviolet sterilizers are often used to help control unwanted microorganisms in aquaria and ponds. UV irradiation ensures that pathogens cannot reproduce, thus decreasing the likelihood of a disease outbreak in an aquarium.

Aquarium and pond sterilizers are typically small, with fittings for tubing that allows the water to flow through the sterilizer on its way from a separate external filter or water pump. Within the sterilizer, water flows as close as possible to the ultraviolet light source. Water pre-filtration is critical as water turbidity lowers UVC penetration. Many of the better UV sterilizers have long dwell times and limit the space between the UVC source and the inside wall of the UV sterilizer device.

Laboratory Hygiene

UVGI is often used to disinfect equipment such as safety goggles, instruments, pipettors, and other devices. Lab personnel also disinfect glassware and plasticware this way. Microbiology laboratories use UVGI to disinfect surfaces inside biological safety cabinets ("hoods") between uses.

Food and Beverage Protection

Since the U.S. Food and Drug Administration issued a rule in 2001 requiring that virtually all fruit and vegetable juice producers follow HACCP controls, and mandating a 5-log reduction in pathogens, UVGI has seen some use in sterilization of juices such as fresh-pressed apple cider.

Technology

Lamps

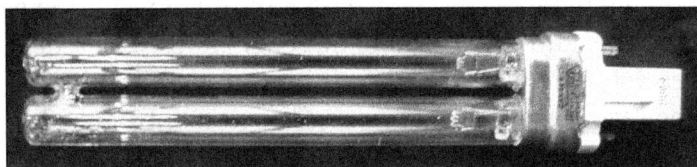

A 9 W germicidal lamp in a compact fluorescent lamp form factor.

Germicidal UV for disinfection is most typically generated by a mercury-vapor lamp. Low-pressure mercury vapor has a strong emission line at 254 nm, which is within the range of wavelengths that demonstrate strong disinfection effect. The optimal wavelengths for disinfection are close to 270 nm.

Mercury vapor amps may be categorized as either low pressure (including amalgam) or medium-pressure lamps. Low-pressure UV lamps offer high efficiencies (approx 35% UVC) but lower power, typically 1 W/cm power density (power per unit of arc length). Amalgam UV lamps utilize an amalgam to control mercury pressure to allow operation at somewhat higher temperature and power density. They operate at higher temperatures and have a lifetime of up to 16,000 hours. Their efficiency is slightly lower than that of traditional low-pressure lamps (approx 33% UVC output) and power density is approximately 2–3 W/cm. Medium-pressure UV lamps operate at much higher temperatures, up to about 800 degrees Celsius, and have a polychromatic output spectrum and a high radiation output but lower UVC efficiency of 10% or less. Typical power density is 30 W/cm³ or greater.

Depending on the quartz glass used for the lamp body, low-pressure and amalgam UV emit radiation at 254 nm and also at 185 nm, which has chemical effects. UV radiation at 185 nm is used to generate ozone.

The UV lamps for water treatment consist of specialized low-pressure mercury-vapor lamps that produce ultraviolet radiation at 254 nm, or medium-pressure UV lamps that produce a polychromatic output from 200 nm to visible and infrared energy. The UV lamp never contacts the water; it is either housed in a quartz glass sleeve inside the water chamber or mounted external to the water which flows through the transparent UV tube. Water passing through the flow chamber is exposed to UV rays which are absorbed by suspended solids, such as microorganisms and dirt, in the stream.

Light Emitting Diodes (LEDs)

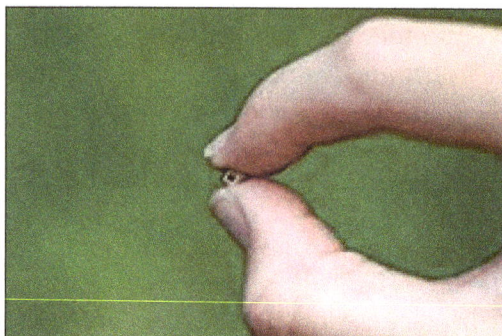

Compact and versatile options with UV-C LEDs.

Recent developments in LED technology have led to commercially available UV-C LEDs. UV-C LEDs use semiconductors to emit light between 255 nm-280 nm. The wavelength emission is tuneable by adjusting the material of the semiconductor. As of 2019, the electrical-to-UVC conversion efficiency of LEDs was lower than that of mercury lamps. The reduced size of LEDs opens up options for small reactor systems allowing for point-of-use applications and integration into medical devices. Low power consumption of semiconductors introduce UV disinfection systems that utilized small solar cells in remote or Third World applications.

UV-C LEDs don't necessarily last longer than traditional germicidal lamps in terms of hours used, and instead have more-variable engineering characteristics and better tolerance for short-term operation. A UV-C LED can achieve a longer installed time than a traditional germicidal lamp in intermittent use. Likewise, LED degradation increases with heat, while filament and HID lamp output wavelength is dependent on temperature, so engineers can design LEDs of particular size and cost to have higher output and faster degradation or lower output and a slower decline over time.

Water Treatment Systems

Sizing of a UV system is affected by three variables: flow rate, lamp power, and UV transmittance in the water. Manufacturers typically developed sophisticated Computational Fluid Dynamics (CFD) models validated with bioassay testing. This involves testing the UV reactor's disinfection performance with either MS2 or T1 bacteriophages at various flow rates, UV transmittance, and power levels in order to develop a regression model for system sizing. For example, this is a requirement for all drinking water systems in the United States per the EPA UV Guidance Manual.

The flow profile is produced from the chamber geometry, flow rate, and particular turbulence model selected. The radiation profile is developed from inputs such as water quality, lamp type (power, germicidal efficiency, spectral output, arc length), and the transmittance and dimension of the quartz sleeve. Proprietary CFD software simulates both the flow and radiation profiles. Once the 3D model of the chamber is built, it is populated with a grid or mesh that comprises thousands of small cubes.

Points of interest—such as at a bend, on the quartz sleeve surface, or around the wiper mechanism—use a higher resolution mesh, whilst other areas within the reactor use a coarse mesh. Once the mesh is produced, hundreds of thousands of virtual particles are "fired" through the chamber. Each particle has several variables of interest associated with it, and the particles are "harvested" after the reactor. Discrete phase modeling produces delivered dose, head loss, and other chamber-specific parameters.

When the modeling phase is complete, selected systems are validated using a professional third party to provide oversight and to determine how closely the model is able to predict the reality of system performance. System validation uses non-pathogenic surrogates such as MS 2 phage or *Bacillus subtilis* to determine the Reduction Equivalent Dose (RED) ability of the reactors. Most systems are validated to deliver 40 mJ/cm^2 within an envelope of flow and transmittance.

To validate effectiveness in drinking-water systems, the method described in the EPA UV Guidance Manual is typically used by the U.S., whilst Europe has adopted Germany's DVGW 294 standard. For wastewater systems, the NWRI/AwwaRF Ultraviolet Disinfection Guidelines for Drinking Water and Water Reuse protocols are typically used, especially in wastewater reuse applications.

UV Degradation

Many natural and synthetic polymers are attacked by ultraviolet radiation, and products using these materials may crack or disintegrate if they are not UV-stable. The problem is known as *UV degradation*, and is a common problem in products exposed to sunlight. Continuous exposure is a more serious problem than intermittent exposure, since attack is dependent on the extent and degree of exposure.

Many pigments and dyes can also be affected, and the problem known as phototendering can affect textiles such as curtains or drapes.

Susceptible Polymers

Common synthetic polymers that can be attacked include polypropylene and LDPE, where tertiary carbon bonds in their chain structures are the centres of attack. Ultraviolet rays interact with these bonds to form free radicals, which then react further with oxygen in the atmosphere, producing carbonyl groups in the main chain. The exposed surfaces of products may then discolour and crack, and in extreme cases, complete product disintegration can occur.

Effect of UV exposure on polypropylene rope.

In fibre products like rope used in outdoor applications, product life will be low because the outer fibres will be attacked first, and will easily be damaged by abrasion for example. Discolouration of the rope may also occur, thus giving an early warning of the problem.

Polymers which possess UV-absorbing groups such as aromatic rings may also be sensitive to UV degradation. Aramid fibres like Kevlar, for example, are highly UV-sensitive and must be protected from the deleterious effects of sunlight.

Detection

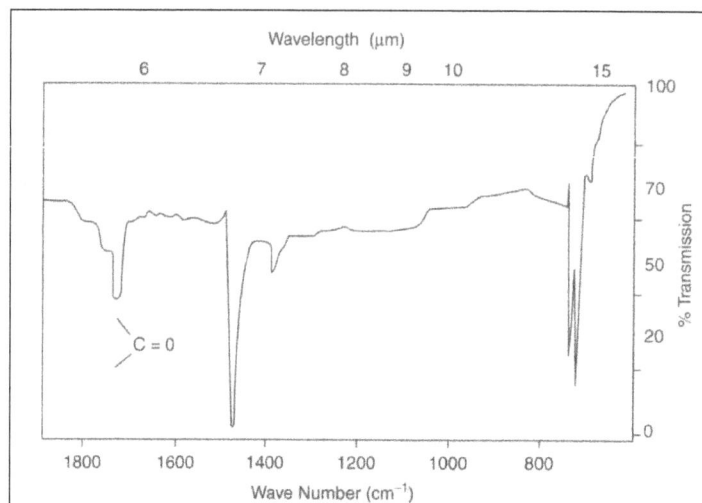

IR spectrum showing carbonyl absorption due to UV degradation of polyethylene.

The problem can be detected before serious cracks are seen in a product using infrared spectroscopy, where attack occurs by oxidation of bonds activated by the UV radiation forming carbonyl groups in the polymer chains.

In the example shown at left, carbonyl groups were easily detected by IR spectroscopy from a cast thin film. The product was a road cone made by rotational moulding in LDPE, which had cracked prematurely in service. Many similar cones also failed because an anti-UV additive had not been used during processing. Other plastic products which failed included polypropylene mancabs used at roadworks which cracked after service of only a few months.

Prevention

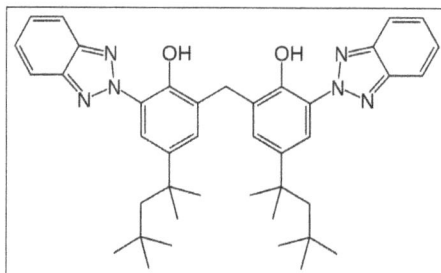

Bisoctrizole: A benzotriazole-phenol based UV absorber used to protect polymers.

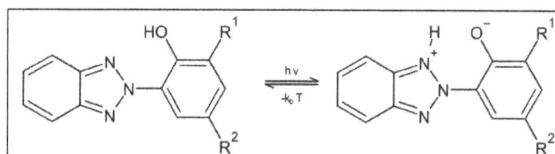

Active principle of the ultraviolet absorption via a photochromic transition.

UV attack by sunlight can be ameliorated or prevented by adding anti-UV chemicals to the polymer when mixing the ingredients, prior to shaping the product by injection moulding for example. UV stabilizers in plastics usually act by absorbing the UV radiation preferentially, and dissipating the energy as low-level heat. The chemicals used are similar to those in sunscreen products, which protect skin from UV attack. They are used frequently in plastics, including cosmetics and films. Different UV stabilizers are utilized depending upon the substrate, intended functional life, and sensitivity to UV degradation. UV stabilizers, such as benzophenones, work by absorbing the UV radiation and preventing the formation of free radicals. Depending upon substitution, the UV absorption spectrum is changed to match the application. Concentrations normally range from 0.05% to 2%, with some applications up to 5%.

Frequently, glass can be a better alternative to polymers when it comes to UV degradation. Most of the commonly used glass types are highly resistant to UV radiation. Explosion protection lamps for oil rigs for example can be made either from polymer or glass. Here, the UV radiation and rough weathers belabor the polymer so much, that the material has to be replaced frequently.

Materials Testing

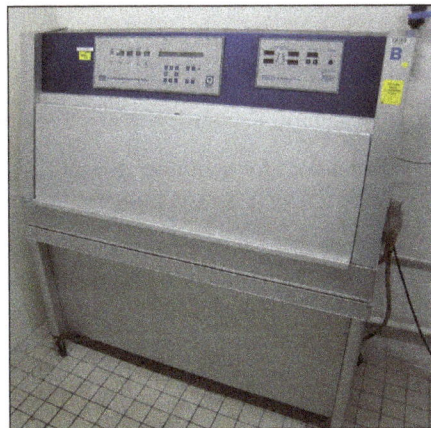

Example of test device.

The effects of UV degradation on materials that require a long service life can be measured with accelerated exposure tests. With modern solar concentrator technologies, it is possible to simulate 63 years of natural UV radiation exposure on a test device in a single year.

UV Pinning

UV pinning is the process of applying a dose of low intensity ultraviolet (UV) light to a UV curable ink (UV ink). The light's wavelengths must be correctly matched to the ink's photochemical properties. As a result, the ink droplets move to a higher viscosity state, but stop short of full cure. This is also referred to as the "gelling" of the ink.

UV pinning is typically used in UV ink jet applications (e.g. the printing of labels, the printing of electronics, and the fabrication of 3-D microstructures).

Purpose

UV pinning enhances the management of drop size and image integrity, minimizing the unwanted mixing of drops and providing the highest possible image quality and the sharpest colour rendering.

Challenge: Overcome the wetting problems that were causing UV-Curable inks to spread and cause ink droplets to bleed into each other before full curing single-pass digital printing of narrow web labels.

Solution: A UV pinning system that uses high power UV light emitting diodes(LEDs) installed next to the inkjet array (print head). The UV light from the pinning system, typically lower than that of the full cure UV system, causes the UV ink to thicken, also known as gelling, but not fully cure. This ink thickening stops dot gain and holds the ink droplet pattern in place until it reaches the full cure UV system.

References

- Bolton, james; colton, christine (2008). The ultraviolet disinfection handbook. American water works association. Pp. 3–4. Isbn 978-1-58321-584-5

- Ultraviolet, entry: newworldencyclopedia.org, Retrieved 18 April, 2019

- "the nobel prize in physiology or medicine 1903". Nobelprize.org. The nobel foundation. Retrieved 2006-09-09

- Wei deng; qi luo (2012). Advanced technology for manufacturing systems and industry. Trans tech publications ltd. Pp. 771–. Isbn 978-3-03813-912-6

- Somiya, shigeyuki, ed. (2003). Handbook of advanced ceramics: materials, applications, processing, and properties (2d ed.). Academic press. Isbn 978-0-12-385469-8. Retrieved 2018-01-21 – via google books

5

Diverse Aspects of Photochemistry

Some of the vital aspects of photochemistry are photocatalysis, photoelectrochemical processes, photovoltaic effect, photovoltaic catalysis and photoelectric effect, etc. It also includes various types of photovoltaic cells such as thin-film solar cell, organic solar cell and hybrid solar cell. This chapter discusses in detail all these diverse aspects of photochemistry.

Photocatalysis

Photocatalysis is a rapidly developing field of research with a high potential for a wide range of industrial applications, which include mineralization of organic pollutants, disinfection of water and air, production of renewable fuels, and organic syntheses. In this process light is used to activate a substance, the photocatalyst, which modifies the rate of a chemical reaction without being involved itself in the chemical transformation. Thus, the main difference between a conventional thermal catalyst and a photocatalyst is that the former is activated by heat whereas the latter is activated by photons of appropriate energy. Photocatalytic reactions may occur homogeneously or heterogeneously, but heterogeneous photocatalysis is by far more intensively studied in recent years because of its potential use in a variety of environmental and energy-related applications as well as in organic syntheses. In heterogeneous photocatalysis, the reaction scheme implies the previous formation of an interface between a solid photocatalyst (metal or semiconductor) and a fluid containing the reactants and products of the reaction. Processes involving illuminated adsorbate-metal interfaces are generally categorized in the branch of photochemistry. Therefore, the term "heterogeneous photocatalysis" is mainly used in cases where a light-absorbing semiconductor photocatalyst is utilized, which is in contact with either a liquid or a gas phase. Here, we will restrict our attention to the description of semiconductor-mediated photocatalysis

Semiconductors are particularly useful as photocatalysts because of a favorable combination of electronic structure, light absorption properties, charge transport characteristics and excited-state lifetimes. A semiconductor, by definition, is nonconductive in its undoped ground state because an energy gap, the bandgap, exists between the top of the filled valence band and the bottom of the vacant conduction band. Thus, electron transport between these bands must occur only with appreciable energy change. In semiconductor photocatalysis, excitation of an electron from the valence band to the conduction band is accomplished by absorption of a photon of energy equal to or higher than the bandgap energy of the semiconductor. This lightinduced generation of an electron-hole pair is a prerequisite step in all semiconductormediated photocatalytic processes. Photogenerated species tend to recombine and dissipate energy as heat of photons, because the kinetic barrier for the electron-hole recombination process is low. However, conduction band electrons and valence band holes can be separated efficiently in the presence of an electric field, such as the one formed spontaneously in the space charge layer of a semiconductor-fluid or a semiconductor-metal interface. Thus, the lifetimes of photogenerated carriers increase and the possibility is offered to these species to exchange charge with substrates adsorbed on the photocatalyst surface and initiate chemical reactions.

Interfacial electron transfer, i.e., transfer of an electron to or from a substrate adsorbed onto the light-activated semiconductor is probably the most critical step in photocatalytic processes, and its efficiency determines to a large extent the ability of the semiconductor to serve as a photocatalyst for a given redox reaction. The efficiency of electron transfer reactions is, in turn, a function of the position of semiconductor's conduction and valence band-edges relative to the redox potentials of the adsorbed substrates. For a desired electron transfer reaction to occur, the potential of the electron acceptor species should be located below (more positive than) the conduction band of the semiconductor, whereas the potential of the electron donor species should be located above (more negative than) the valence band of the semiconductor. Interfacial electron transfer processes are then initiating subsequent (dark) redox reactions to yield the ultimate products. The latter catalytic reaction steps are not different from thermal catalysis, and similar principles will apply for a photocatalytic reaction as for a thermal catalytic reaction.

The overall scheme of a semiconductor-mediated photocatalytic reaction is often described by the following general equation:

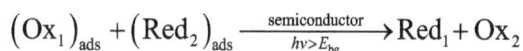

$$\left(Ox_1\right)_{ads} + \left(Red_2\right)_{ads} \xrightarrow[h\nu > E_{bg}]{semiconductor} Red_1 + Ox_2$$

The potential energy required for the chemical transformation to occur is overcome by the large amounts of "free energy" supplied with the ultra-violet or visible light quanta. Thus, in contrast to conventional thermal catalytic reactions, which usually rely on the application of high temperature and/or pressure, photocatalysis occurs under relatively mild (usually ambient) conditions with lower energy input. This is particularly true when the photocatalytic reaction is activated by solar light. An important characteristic of photocatalysts is that, in contrast to thermal catalysts, they can drive thermodynamically "uphill" reactions, i.e., reactions which involve transformations of substrates to products of higher potential energy. In this respect, a distinction is often made between "photocatalytic" and "photosynthetic" reactions, depending on whether the sign of the change in Gibbs free energy $\left(\Delta G°\right)$ of reaction (1) is negative or positive, respectively. Examples of spontaneous, exergonic reactions $\left(\Delta G° < 0\right)$ include oxidation of organic pollutants present in the gas or liquid phase, whereas an example of a photosynthetic, endergonic reaction $\left(\Delta G° > 0\right)$ is the photocatalytic cleavage of water toward hydrogen $\left(H_2\right)$ and oxygen $\left(O_2\right)$.

A large variety of semiconducting materials, mainly metal oxides and chalcogenides, have been investigated with respect to their photocatalytic properties, but only few of them are considered to be effective photocatalysts. In general, wide-bandgap semiconductors, such as titanium dioxide (TiO_2), prove to be better photocatalysts than low-bandgap materials, such as cadmium sulfide (CdS), mainly due to the higher free energy of photogenerated charge carriers of the former and the inherently low chemical and photochemical stability of the latter. However, low bandgap semiconductors are better adapted to the solar spectrum, thereby offering the significant advantage of potential utilization of a continuous and readily available power supply, the sun. A considerable amount of effort has been made in recent years for the development of more efficient photocatalysts characterized by increased quantum efficiency and improved response to the visible spectral region. Promising results in this direction have been obtained with the use of several methods aiming at the modification of electronic and/or optical properties of semiconductors, including metal deposition, dye sensitization, doping with transition metals or non-metallic elements, use of composite semiconductor photocatalysts, etc.

The potential applications of heterogeneous photocatalysis depend strongly on the development of scaled-up reactor designs with increased efficiency. The major challenge in the design of a photocatalytic reactor is the efficient illumination of the catalyst and mass transfer optimization, especially in liquid phase reactions. Mass transfer limitations can be dealt with by the use of, for example, spinning disc reactors, monolithic reactors and microreactors, which have been proven to be much more efficient than conventional reactors. Photon transfer can be optimized with the use of optical fibers and light-emitting diodes (LEDs), but major breakthroughs in this field are still lacking. Since the artificial generation of photons required for photocatalytic reactions is the most important source of operating costs in practical applications, a significant amount of research effort has been directed toward the development of solar photoreactors. Among the various types of solar reactor configurations evaluated so far, compound parabolic collectors are the most promising, and have been successfully scaled-up for applications related to wastewater treatment and water cleaning and disinfection.

Semiconductor photocatalysis is currently one of the most active interdisciplinary research areas, and has been investigated from the standpoint of catalysis, photochemistry, electrochemistry, inorganic, organic, physical, polymer and environmental chemistry. As a result of numerous investigations in these fields, the fundamental processes of photocatalysis are now much better understood. The applicability of photocatalysis has been proven in laboratory scale for a great number of different processes ranging from water treatment and air cleaning to disinfection and anti-tumoral applications, and from production of fuels from water and atmospheric gases to selective organic synthesis and metal recovery. However, industrial applications still remain limited. The current lack of widespread industrial applications is mainly due to the low photocatalytic efficiency of semiconductor photocatalysts and the absence of efficient, large scale photoreactor configurations.

Photoelectrochemical Process

Photoelectrochemical processes are processes in photoelectrochemistry; they usually involve transforming light into other forms of energy. These processes apply to photochemistry, optically pumped lasers, sensitized solar cells, luminescence, and photochromism.

Electron Excitation

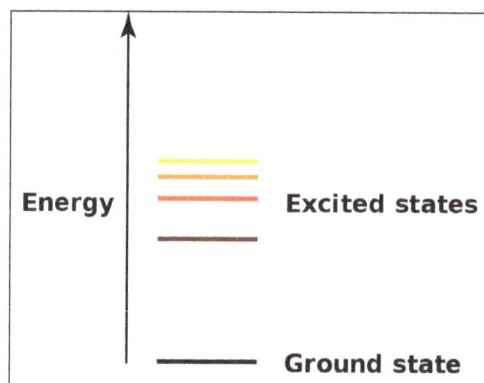

After absorbing energy, an electron may jump
from the ground state to a higher energy excited state.

Electron excitation is the movement of an electron to a higher energy state. This can either be done by photoexcitation (PE), where the original electron absorbs the photon and gains all the photon's energy or by electrical excitation (EE), where the original electron absorbs the energy of another, energetic electron. Within a semiconductor crystal lattice, thermal excitation is a process where lattice vibrations provide enough energy to move electrons to a higher energy band. When an excited electron falls back to a lower energy state again, it is called electron relaxation. This can be done by radiation of a photon or giving the energy to a third spectator particle as well.

In physics there is a specific technical definition for energy level which is often associated with an atom being excited to an excited state. The excited state, in general, is in relation to the ground state, where the excited state is at a higher energy level than the ground state.

Photoexcitation

Photoexcitation is the mechanism of electron excitation by photon absorption, when the energy of the photon is too low to cause photoionization. The absorption of the photon takes place in accordance with Planck's quantum theory.

Photoexcitation plays role in photoisomerization. Photoexcitation is exploited in dye-sensitized solar cells, photochemistry, luminescence, optically pumped lasers, and in some photochromic applications.

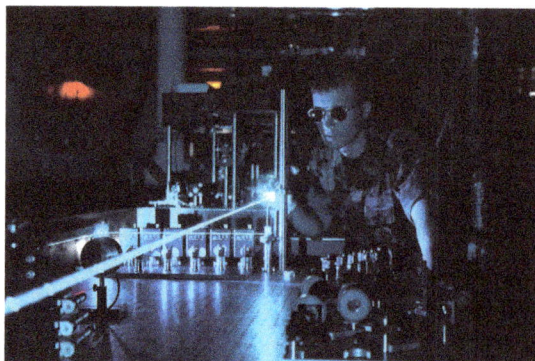

Military laser experiment.

Photoisomerization

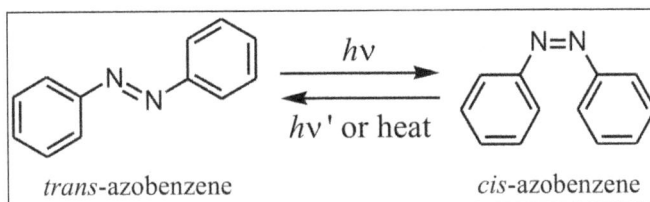

Photoisomerization of azobenzene.

In chemistry, photoisomerization is molecular behavior in which structural change between isomers is caused by photoexcitation. Both reversible and irreversible photoisomerization reactions exist. However, the word "photoisomerization" usually indicates a reversible process. Photoisomerizable molecules are already put to practical use, for instance, in pigments for rewritable CDs, DVDs, and 3D optical data storage solutions. In addition, recent interest in photoisomerizable molecules has been aimed at molecular devices, such as molecular switches, molecular motors, and molecular electronics.

Photoisomerization behavior can be roughly categorized into several classes. Two major classes are *trans-cis* (or 'E-'Z) conversion, and open-closed ring transition. Examples of the former include stilbene and azobenzene. This type of compounds has a double bond, and rotation or inversion around the double bond affords isomerization between the two states. Examples of the latter include fulgide and diarylethene. This type of compounds undergoes bond cleavage and bond creation upon irradiation with particular wavelengths of light. Still another class is the Di-pi-methane rearrangement.

Photoionization

Photoionization is the physical process in which an incident photon ejects one or more electrons from an atom, ion or molecule. This is essentially the same process that occurs with the photoelectric effect with metals. In the case of a gas or single atoms, the term photoionization is more common.

The ejected electrons, known as photoelectrons, carry information about their pre-ionized states. For example, a single electron can have a kinetic energy equal to the energy of the incident photon minus the electron binding energy of the state it left. Photons with energies less than the electron binding energy may be absorbed or scattered but will not photoionize the atom or ion.

For example, to ionize hydrogen, photons need an energy greater than 13.6 electronvolts (the Rydberg energy), which corresponds to a wavelength of 91.2 nm. For photons with greater energy than this, the energy of the emitted photoelectron is given by:

$$\frac{mv^2}{2} = h\nu - 13.6 eV$$

where h is Planck's constant and ν is the frequency of the photon.

This formula defines the photoelectric effect.

Not every photon which encounters an atom or ion will photoionize it. The probability of photoionization is related to the photoionization cross-section, which depends on the energy of the photon and the target being considered. For photon energies below the ionization threshold, the photoionization cross-section is near zero. But with the development of pulsed lasers it has become possible to create extremely intense, coherent light where multi-photon ionization may occur. At even higher intensities (around 10^{15} - 10^{16} W/cm^2 of infrared or visible light), non-perturbative phenomena such as *barrier suppression ionization* and *rescattering ionization* are observed.

Multi-photon Ionization

Several photons of energy below the ionization threshold may actually combine their energies to ionize an atom. This probability decreases rapidly with the number of photons required, but the development of very intense, pulsed lasers still makes it possible. In the perturbative regime (below about 10^{14} W/cm^2 at optical frequencies), the probability of absorbing N photons depends on the laser-light intensity I as I^N.

Above threshold ionization (ATI) is an extension of multi-photon ionization where even more photons are absorbed than actually would be necessary to ionize the atom. The excess energy gives the released electron higher kinetic energy than the usual case of just-above threshold ionization. More precisely, the system will have multiple peaks in its photoelectron spectrum which are separated by the photon energies, this indicates that the emitted electron has more kinetic energy than in the normal (lowest possible number of photons) ionization case. The electrons released from the target will have approximately an integer number of photon-energies more kinetic energy. In intensity regions between 10^{14} W/cm^2 and 10^{18} W/cm^2, each of MPI, ATI, and barrier suppression ionization can occur simultaneously, each contributing to the overall ionization of the atoms involved.

Photo-Dember

In semiconductor physics the Photo-Dember effect (named after its discoverer H. Dember) consists in the formation of a charge dipole in the vicinity of a semiconductor surface after ultra-fast photo-generation of charge carriers. The dipole forms owing to the difference of mobilities (or diffusion constants) for holes and electrons which combined with the break of symmetry provided by the surface lead to an effective charge separation in the direction perpendicular to the surface.

Grotthuss–Draper Law

The Grotthuss–Draper law (also called the Principle of Photochemical Activation) states that only that light which is absorbed by a system can bring about a photochemical change. Materials such as dyes and phosphors must be able to absorb "light" at optical frequencies. This law provides a basis for fluorescence and phosphorescence. The law was first proposed in 1817 by Theodor Grotthuss and in 1842, independently, by John William Draper.

This is considered to be one of the two basic laws of photochemistry. The second law is the Stark–Einstein law, which says that primary chemical or physical reactions occur with each photon absorbed.

Stark–Einstein law

The Stark–Einstein law is named after German-born physicists Johannes Stark and Albert Einstein, who independently formulated the law between 1908 and 1913. It is also known as the photochemical equivalence law or photoequivalence law. In essence it says that every photon that is absorbed will cause a (primary) chemical or physical reaction.

The photon is a quantum of radiation, or one unit of radiation. Therefore, this is a single unit of EM radiation that is equal to Planck's constant (h) times the frequency of light. This quantity is symbolized by γ, $h\nu$, or $\hbar\omega$.

The photochemical equivalence law is also restated as follows: for every mole of a substance that reacts, an equivalent mole of quanta of light are absorbed. The formula is:

$$\Delta E_{mol} = N_A h\nu$$

where N_A is Avogadro's number.

The photochemical equivalence law applies to the part of a light-induced reaction that is referred to as the primary process (i.e. absorption or fluorescence).

In most photochemical reactions the primary process is usually followed by so-called secondary photochemical processes that are normal interactions between reactants not requiring absorption of light. As a result, such reactions do not appear to obey the one quantum–one molecule reactant relationship.

The law is further restricted to conventional photochemical processes using light sources with moderate intensities; high-intensity light sources such as those used in flash photolysis and in laser experiments are known to cause so-called biphotonic processes; i.e., the absorption by a molecule of a substance of two photons of light.

Absorption

In physics, absorption of electromagnetic radiation is the way by which the energy of a photon is taken up by matter, typically the electrons of an atom. Thus, the electromagnetic energy is transformed to other forms of energy, for example, to heat. The absorption of light during wave propagation is often called attenuation. Usually, the absorption of waves does not depend on their intensity (linear absorption), although in certain conditions (usually, in optics), the medium changes its transparency dependently on the intensity of waves going through, and the Saturable absorption (or nonlinear absorption) occurs.

Photosensitization

Photosensitization is a process of transferring the energy of absorbed light. After absorption, the energy is transferred to the (chosen) reactants. This is part of the work of photochemistry in general. In particular this process is commonly employed where reactions require light sources of certain wavelengths that are not readily available.

For example, mercury absorbs radiation at 1849 and 2537 angstroms, and the source is often high-intensity mercury lamps. It is a commonly used sensitizer. When mercury vapor is mixed with ethylene, and the compound is irradiated with a mercury lamp, this results in the photodecomposition of ethylene to acetylene. This occurs on absorption of light to yield excited state mercury atoms, which are able to transfer this energy to the ethylene molecules, and are in turn deactivated to their initial energy state.

Cadmium; some of the noble gases, for example xenon; zinc; benzophenone; and a large number of organic dyes, are also used as sensitizers.

Photosensitisers are a key component of photodynamic therapy used to treat cancers.

Sensitizer

A sensitizer in chemiluminescence is a chemical compound, capable of light emission after it has received energy from a molecule, which became excited previously in the chemical reaction. A good example is this:

When an alkaline solution of sodium hypochlorite and a concentrated solution of hydrogen peroxide are mixed, a reaction occurs:

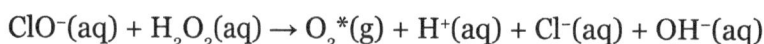

$$ClO^-(aq) + H_2O_2(aq) \rightarrow O_2^*(g) + H^+(aq) + Cl^-(aq) + OH^-(aq)$$

O_2^* is excited oxygen – meaning, one or more electrons in the O_2 molecule have been promoted to higher-energy molecular orbitals. Hence, oxygen produced by this chemical reaction somehow 'absorbed' the energy released by the reaction and became excited. This energy state is unstable, therefore it will return to the ground state by lowering its energy. It can do that in more than one way:

- It can react further, without any light emission.

- It can lose energy without emission, for example, giving off heat to the surroundings or transferring energy to another molecule.

- It can emit light.

The intensity, duration and color of emitted light depend on quantum and kinetical factors. However, excited molecules are frequently less capable of light emission in terms of brightness and duration when compared to sensitizers. This is because sensitizers can store energy (that is, be excited) for longer periods of time than other excited molecules. The energy is stored through means of quantum vibration, so sensitizers are usually compounds which either include systems of aromatic rings or many conjugated double and triple bonds in their structure. Hence, if an excited molecule transfers its energy to a sensitizer thus exciting it, longer and easier to quantify light emission is often observed.

The color (that is, the wavelength), brightness and duration of emission depend upon the sensitizer used. Usually, for a certain chemical reaction, many different sensitizers can be used.

List of Some Common Sensitizers

- Violanthrone,

- Isoviolanthrone,

- Fluorescein,

- Rubrene,

- 9,10-Diphenylanthracene,

- Tetracene,

- 13,13'-Dibenzantronile,

- Levulinic Acid.

Fluorescence Spectroscopy

Fluorescence spectroscopy aka fluorometry or spectrofluorometry, is a type of electromagnetic spectroscopy which analyzes fluorescence from a sample. It involves using a beam of light, usually ultraviolet light, that excites the electrons in molecules of certain compounds and causes them to emit light of a lower energy, typically, but not necessarily, visible light. A complementary technique is absorption spectroscopy.

Devices that measure fluorescence are called fluorometers or fluorimeters.

Absorption Spectroscopy

Absorption spectroscopy refers to spectroscopic techniques that measure the absorption of radiation, as a function of frequency or wavelength, due to its interaction with a sample. The sample absorbs energy, i.e., photons, from the radiating field. The intensity of the absorption varies as a function of frequency, and this variation is the absorption spectrum. Absorption spectroscopy is performed across the electromagnetic spectrum.

Photoelectrochemical Cell

A "photoelectrochemical cell" is one of two distinct classes of device. The first produces electrical energy similarly to a dye-sensitized photovoltaic cell, which meets the standard definition of a photovoltaic cell. The second is a photoelectrolytic cell, that is, a device which uses light incident on a photosensitizer, semiconductor, or aqueous metal immersed in an electrolytic solution to directly cause a chemical reaction, for example to produce hydrogen via the electrolysis of water.

Both types of device are varieties of solar cell, in that a photoelectrochemical cell's function is to use the photoelectric effect (or, very similarly, the photovoltaic effect) to convert electromagnetic radiation (typically sunlight) either directly into electrical power, or into something which can itself be easily used to produce electrical power (hydrogen, for example, can be burned to create electrical power,).

Two Principles

The standard photovoltaic effect, as operating in standard photovoltaic cells, involves the excitation of negative charge carriers (electrons) within a semiconductor medium, and it is negative charge carriers (free electrons) which are ultimately are extracted to produce power. The classification of photoelectrochemical cells which includes Grätzel cells meets this narrow definition, albeit the charge carriers are often excitonic.

The situation within a photoelectrolytic cell, on the other hand, is quite different. For example, in a water-splitting photoelectrochemical cell, the excitation, by light, of an electron in a semiconductor leaves a hole which "draws" an electron from a neighboring water molecule:

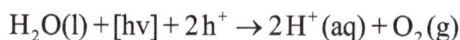

$$H_2O(l) + [hv] + 2h^+ \rightarrow 2H^+(aq) + O_2(g)$$

This leaves positive charge carriers (protons, that is, H+ ions) in solution, which must then bond with one other proton and combine with two electrons in order to form hydrogen gas, according to:

$$2H^+ + 2e \rightarrow H_2(g)$$

A photosynthetic cell is another form of photoelectrolytic cell, with the output in that case being carbohydrates instead of molecular hydrogen.

Photoelectrolytic Cell

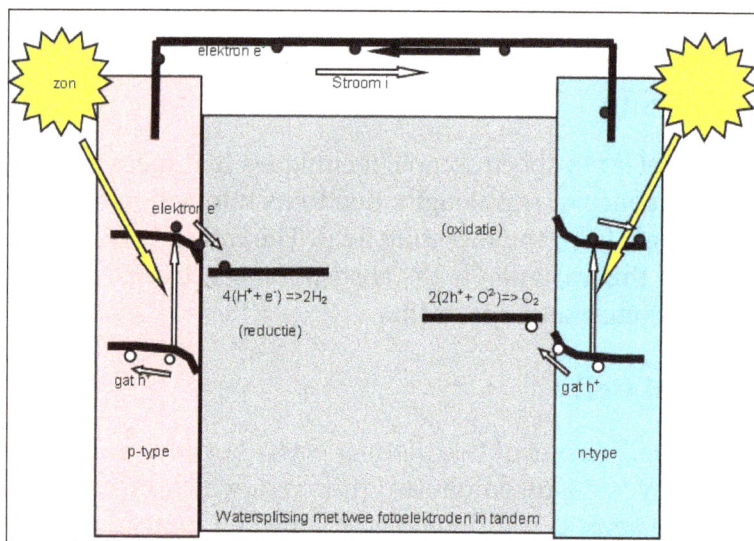

Photoelectrolytic cell band diagram.

A (water-splitting) photoelectrolytic cell electrolizes water into hydrogen and oxygen gas by irradiating the anode with electromagnetic radiation, that is, with light. This has been referred to as artificial photosynthesis and has been suggested as a way of storing solar energy in hydrogen for use as fuel.

Incoming sunlight excites free electrons near the surface of the silicon electrode. These electrons flow through wires to the stainless steel electrode, where four of them react with four water molecules to form two molecules of hydrogen and 4 OH groups. The OH groups flow through the liquid electrolyte to the surface of the silicon electrode. There they react with the four holes associated with the four photoelectrons, the result being two water molecules and an oxygen molecule. Illuminated silicon immediately begins to corrode under contact with the electrolytes. The corrosion consumes material and disrupts the properties of the surfaces and interfaces within the cell.

Two types of photochemical systems operate via photocatalysis. One uses semiconductor surfaces as catalysts. In these devices the semiconductor surface absorbs solar energy and acts as an electrode for water splitting. The other methodology uses in-solution metal complexes as catalysts.

Photoelectrolytic cells have passed the 10 percent economic efficiency barrier. Corrosion of the semiconductors remains an issue, given their direct contact with water. Research is now ongoing to reach a service life of 10000 hours, a requirement established by the United States Department of Energy.

Other Photoelectrochemical Cells

The first photovoltaic cell ever designed was also the first photoelectrochemical cell. It was created in 1839, by Alexandre-Edmond Becquerel, at age 19, in his father's laboratory.

The mostly commonly researched modern photoelectrochemical cell in recent decades has been the Grätzel cell, although much attention has recently shifted away from this topic to perovskite solar cells, due to relatively high efficiency of the latter and the similarity in vapor assisted deposition techniques commonly used in their creation.

Dye-sensitized solar cells or Grätzel cells use dye-adsorbed highly porous nanocrystalline titanium dioxide (nc-TiO_2) to produce electrical energy.

Materials for Photoelectrolytic Cells

Water-splitting photoelectrolytic photoelectrochemical cells (PECs) use light energy to extract hydrogen from water within a two-electrode cell. In theory, three arrangements of photo-electrodes in the assembly of PECs exist:

- Photo-anode made of a n-type semiconductor and a metal cathode.

- Photo-anode made of a n-type semiconductor and a photo-cathode made of a p-type semiconductor.

- Photo-cathode made of a p-type semiconductor and a metal anode.

TiO_2

In 1967, Akira Fujishima discovered the Honda-Fujishima effect, (the photocatalytic properties of titanium dioxide).

TiO_2 and other metal oxides are still most prominent catalysts for efficiency reasons. Including $SrTiO_3$ and $BaTiO_3$, this kind of semiconducting titanates, the conduction band has mainly titanium 3d character and the valence band oxygen 2p character. The bands are separated by a wide band gap of at least 3 eV, so that these materials absorb only UV radiation. Change of the TiO_2 microstructure has also been investigated to further improve the performance, such as TiO_2 nanowire arrays or porous nanocrystalline TiO_2 photoelectrochemical cells.

GaN

GaN is another option, because metal nitrides usually have a narrow band gap that could encompass almost the entire solar spectrum. GaN has a narrower band gap than TiO_2 but is still large enough to allow water splitting to occur at the surface. GaN nanowires exhibited better performance than GaN thin films, because they have a larger surface area and have a high single crystallinity which allows longer electron-hole pair lifetimes. Meanwhile, other non-oxide semiconductors such as

GaAs, MoS_2, WSe_2 and $MoSe_2$ are used as n-type electrode, due to their stability in chemical and electrochemical steps in the photocorrosion reactions.

Silicon

In 2013 a cell with 2 nanometers of nickel on a silicon electrode, paired with a stainless steel electrode, immersed in an aqueous electrolyte of potassium borate and lithium borate operated for 80 hours without noticeable corrosion, versus 8 hours for titanium dioxide. In the process, about 150 ml of hydrogen gas was generated, representing the storage of about 2 kilojoules of energy.

Photovoltaic Effect

The photovoltaic effect is the creation of voltage and electric current in a material upon exposure to light. It is a physical and chemical phenomenon.

The photovoltaic effect is closely related to the photoelectric effect. In either case, light is absorbed, causing excitation of an electron or other charge carrier to a higher-energy state. The main distinction is that the term *photoelectric effect* is now usually used when the electron is ejected out of the material (usually into a vacuum) and *photovoltaic effect* used when the excited charge carrier is still contained within the material. In either case, an electric potential (or voltage) is produced by the separation of charges, and the light has to have a sufficient energy to overcome the potential barrier for excitation. The physical essence of the difference is usually that photoelectric emission separates the charges by ballistic conduction and photovoltaic emission separates them by diffusion, but some "hot carrier" photovoltaic device concepts blur this distinction.

The first demonstration of the photovoltaic effect, by Edmond Becquerel in 1839, used an electrochemical cell. The first solar cell, consisting of a layer of selenium covered with a thin film of gold, was experimented by Charles Fritts in 1884, but it had a very poor efficiency. However, the most familiar form of the photovoltaic effect uses solid-state devices, mainly in photodiodes. When sunlight or other sufficiently energetic light is incident upon the photodiode, the electrons present in the valence band absorb energy and, being excited, jump to the conduction band and become free. These excited electrons diffuse, and some reach the rectifying junction (usually a diode p-n junction) where they are accelerated into the p-type semiconductor material by the built-in potential (Galvani potential). This generates an electromotive force and an electrical current, and thus some of the light energy is converted into electric energy. The photovoltaic effect can also occur when two photons are absorbed simultaneously in a process called two-photon photovoltaic effect.

The photovoltaic effect was first observed by French physicist A. E. Becquerel in 1839. He explained his discovery in *Comptes rendus de l'Académie des sciences*, "the production of an electric current when two plates of platinum or gold immersed in an acid, neutral, or alkaline solution are exposed in an uneven way to solar radiation."

Besides the direct excitation of free electrons, a photovoltaic effect can also arise simply due to the heating caused by absorption of the light. The heating leads to increased temperature of the semiconductor material, which is accompanied by temperature gradients. These thermal gradients in turn may generate a voltage through the Seebeck effect. Whether direct excitation or thermal effects dominate the photovoltaic effect will depend on many material parameters.

In most photovoltaic applications the radiation is sunlight, and the devices are called solar cells. In the case of a semiconductor p-n (diode) junction solar cell, illuminating the material creates an electric current because excited electrons and the remaining holes are swept in different directions by the built-in electric field of the depletion region.

Photovoltaic Cell

A solar cell, or photovoltaic cell, is an electrical device that converts the energy of light directly into electricity by the photovoltaic effect, which is a physical and chemical phenomenon. It is a form of photoelectric cell, defined as a device whose electrical characteristics, such as current, voltage, or resistance, vary when exposed to light. Individual solar cell devices can be combined to form modules, otherwise known as solar panels. The common single junction silicon solar cell can produce a maximum open-circuit voltage of approximately 0.5 to 0.6 volts.

A conventional crystalline silicon solar cell. Electrical contacts
made from busbars (the larger silver-colored strips) and fingers
(the smaller ones) are printed on the silicon wafer.

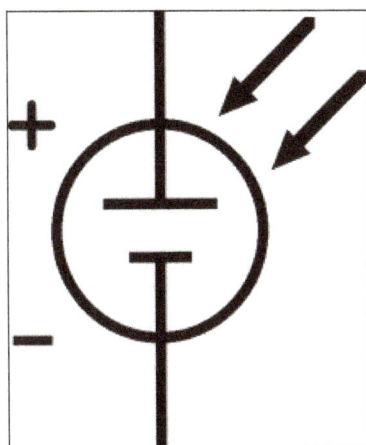

Symbol of a Photovoltaic cell.

Solar cells are described as being photovoltaic, irrespective of whether the source is sunlight or an artificial light. In addition to producing energy, they can be used as a photodetector (for example infrared detectors), detecting light or other electromagnetic radiation near the visible range, or measuring light intensity.

The operation of a photovoltaic (PV) cell requires three basic attributes:

- The absorption of light, generating either electron-hole pairs or excitons.

- The separation of charge carriers of opposite types.

- The separate extraction of those carriers to an external circuit.

In contrast, a solar thermal collector supplies heat by absorbing sunlight, for the purpose of either direct heating or indirect electrical power generation from heat. A "photoelectrolytic cell" (photo-electrochemical cell), on the other hand, refers either to a type of photovoltaic cell (like that developed by Edmond Becquerel and modern dye-sensitized solar cells), or to a device that splits water directly into hydrogen and oxygen using only solar illumination.

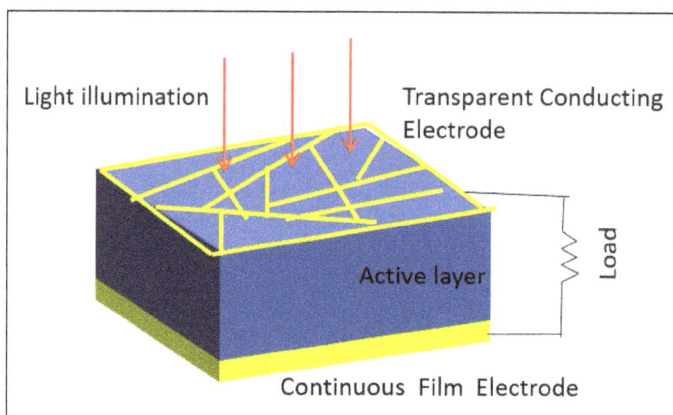

Schematic of charge collection by solar cells. Light transmits through transparent conducting electrode creating electron hole pairs, which are collected by both the electrodes.

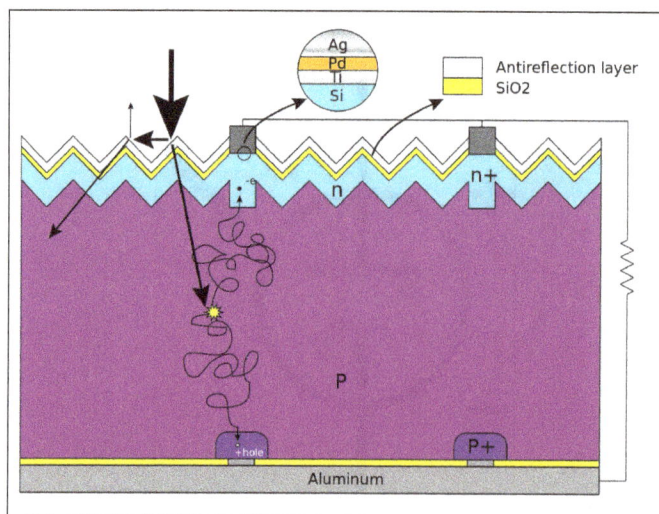

Working mechanism of a solar cell.

The solar cell works in several steps:

- Photons in sunlight hit the solar panel and are absorbed by semiconducting materials, such as silicon.

- Electrons are excited from their current molecular/atomic orbital. Once excited an electron can either dissipate the energy as heat and return to its orbital or travel through the cell until it reaches an electrode. Current flows through the material to cancel the potential and this electricity is captured. The chemical bonds of the material are vital for this process to work, and usually silicon is used in two layers, one layer being doped with boron, the other phosphorus. These layers have different chemical electric charges and subsequently both drive and direct the current of electrons.

- An array of solar cells converts solar energy into a usable amount of direct current (DC) electricity.

- An inverter can convert the power to alternating current (AC).

The most commonly known solar cell is configured as a large-area p–n junction made from silicon. Other possible solar cell types are organic solar cells, dye sensitized solar cells, perovskite solar cells, quantum dot solar cells etc. The illuminated side of a solar cell generally has a transparent conducting film for allowing light to enter into active material and to collect the generated charge carriers. Typically, films with high transmittance and high electrical conductance such as indium tin oxide, conducting polymers or conducting nanowire networks are used for the purpose.

Efficiency

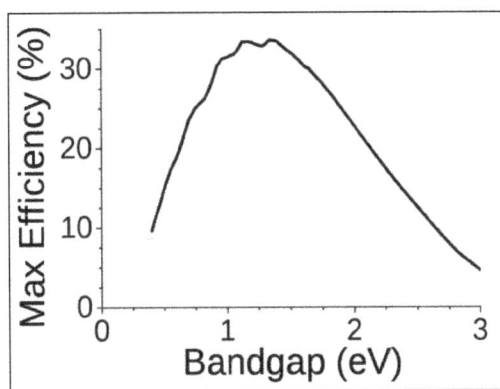

The Shockley-Queisser limit for the theoretical maximum efficiency of a solar cell. Semiconductors with band gap between 1 and 1.5eV, or near-infrared light, have the greatest potential to form an efficient single-junction cell. (The efficiency "limit" shown here can be exceeded by multijunction solar cells.)

Solar cell efficiency may be broken down into reflectance efficiency, thermodynamic efficiency, charge carrier separation efficiency and conductive efficiency. The overall efficiency is the product of these individual metrics.

The power conversion efficiency of a solar cell is a parameter which is defined by the fraction of incident power converted into electricity.

A solar cell has a voltage dependent efficiency curve, temperature coefficients, and allowable shadow angles.

Due to the difficulty in measuring these parameters directly, other parameters are substituted: thermodynamic efficiency, quantum efficiency, integrated quantum efficiency, V_{OC} ratio, and fill factor. Reflectance losses are a portion of quantum efficiency under "external quantum efficiency". Recombination losses make up another portion of quantum efficiency, V_{OC} ratio, and fill factor. Resistive losses are predominantly categorized under fill factor, but also make up minor portions of quantum efficiency, V_{OC} ratio.

The fill factor is the ratio of the actual maximum obtainable power to the product of the open circuit voltage and short circuit current. This is a key parameter in evaluating performance. In 2009, typical commercial solar cells had a fill factor > 0.70. Grade B cells were usually between 0.4 and 0.7. Cells with a high fill factor have a low equivalent series resistance and a high equivalent shunt resistance, so less of the current produced by the cell is dissipated in internal losses.

Single p–n junction crystalline silicon devices are now approaching the theoretical limiting power efficiency of 33.16%, noted as the Shockley–Queisser limit in 1961. In the extreme, with an infinite number of layers, the corresponding limit is 86% using concentrated sunlight.

In 2014, three companies broke the record of 25.6% for a silicon solar cell. Panasonic's was the most efficient. The company moved the front contacts to the rear of the panel, eliminating shaded areas. In addition they applied thin silicon films to the (high quality silicon) wafer's front and back to eliminate defects at or near the wafer surface.

In 2015, a 4-junction GaInP/GaAs//GaInAsP/GaInAs solar cell achieved a new laboratory record efficiency of 46.1% (concentration ratio of sunlight = 312) in a French-German collaboration between the Fraunhofer Institute for Solar Energy Systems (Fraunhofer ISE), CEA-LETI and SOITEC.

In September 2015, Fraunhofer ISE announced the achievement of an efficiency above 20% for epitaxial wafer cells. The work on optimizing the atmospheric-pressure chemical vapor deposition (APCVD) in-line production chain was done in collaboration with NexWafe GmbH, a company spun off from Fraunhofer ISE to commercialize production.

For triple-junction thin-film solar cells, the world record is 13.6%, set in June 2015.

In 2016, researchers at Fraunhofer ISE announced a GaInP/GaAs/Si triple-junction solar cell with two terminals reaching 30.2% efficiency without concentration.

In 2017, a team of researchers at National Renewable Energy Laboratory (NREL), EPFL and CSEM (Switzerland) reported record one-sun efficiencies of 32.8% for dual-junction GaInP/GaAs solar cell devices. In addition, the dual-junction device was mechanically stacked with a Si solar cell, to achieve a record one-sun efficiency of 35.9% for triple-junction solar cells.

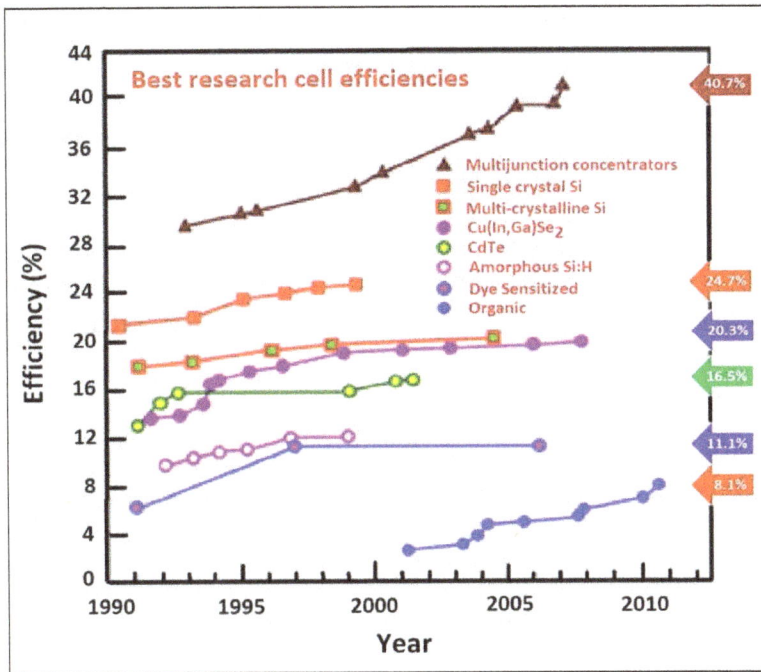

Materials

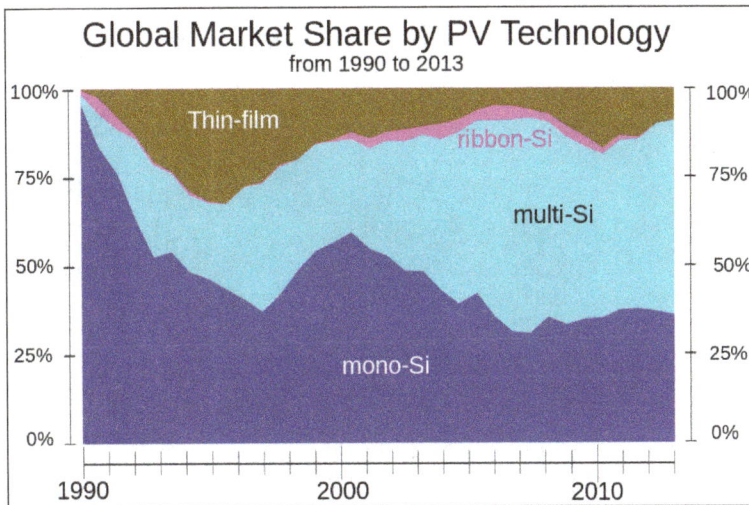

Global market-share in terms of annual production by PV technology since 1990.

Solar cells are typically named after the semiconducting material they are made of. These materials must have certain characteristics in order to absorb sunlight. Some cells are designed to handle sunlight that reaches the Earth's surface, while others are optimized for use in space. Solar cells can be made of only one single layer of light-absorbing material (single-junction) or use multiple physical configurations (multi-junctions) to take advantage of various absorption and charge separation mechanisms.

Solar cells can be classified into first, second and third generation cells. The first generation cells—also called conventional, traditional or wafer-based cells—are made of crystalline silicon, the commercially predominant PV technology, that includes materials such as polysilicon and monocrystalline silicon. Second generation cells are thin film solar cells, that include amorphous silicon, CdTe and CIGS cells and are commercially significant in utility-scale photovoltaic power stations, building integrated photovoltaics or in small stand-alone power system. The third generation of solar cells includes a number of thin-film technologies often described as emerging photovoltaics—most of them have not yet been commercially applied and are still in the research or development phase. Many use organic materials, often organometallic compounds as well as inorganic substances. Despite the fact that their efficiencies had been low and the stability of the absorber material was often too short for commercial applications, there is a lot of research invested into these technologies as they promise to achieve the goal of producing low-cost, high-efficiency solar cells.

Crystalline Silicon

By far, the most prevalent bulk material for solar cells is crystalline silicon (c-Si), also known as "solar grade silicon". Bulk silicon is separated into multiple categories according to crystallinity and crystal size in the resulting ingot, ribbon or wafer. These cells are entirely based around the concept of a p-n junction. Solar cells made of c-Si are made from wafers between 160 and 240 micrometers thick.

Monocrystalline Silicon

The roof, bonnet and large parts of the outer shell of the Sion are
equipped with highly efficient monocrystalline silicon cells

Monocrystalline silicon (mono-Si) solar cells are more efficient and more expensive than most other types of cells. The corners of the cells look clipped, like an octagon, because the wafer material is cut from cylindrical ingots, that are typically grown by the Czochralski process. Solar panels using mono-Si cells display a distinctive pattern of small white diamonds.

Epitaxial Silicon Development

Epitaxial wafers of crystalline silicon can be grown on a monocrystalline silicon "seed" wafer by chemical vapor deposition (CVD), and then detached as self-supporting wafers of some standard thickness (e.g., 250 μm) that can be manipulated by hand, and directly substituted for wafer cells cut from monocrystalline silicon ingots. Solar cells made with this "kerfless" technique can have efficiencies approaching those of wafer-cut cells, but at appreciably lower cost if the CVD can be done at atmospheric pressure in a high-throughput inline process. The surface of epitaxial wafers may be textured to enhance light absorption.

In June 2015, it was reported that heterojunction solar cells grown epitaxially on n-type monocrystalline silicon wafers had reached an efficiency of 22.5% over a total cell area of 243.4 cm.

Polycrystalline Silicon

Polycrystalline silicon, or multicrystalline silicon (multi-Si) cells are made from cast square ingots—large blocks of molten silicon carefully cooled and solidified. They consist of small crystals giving the material its typical metal flake effect. Polysilicon cells are the most common type used in photovoltaics and are less expensive, but also less efficient, than those made from monocrystalline silicon.

Ribbon Silicon

Ribbon silicon is a type of polycrystalline silicon—it is formed by drawing flat thin films from molten silicon and results in a polycrystalline structure. These cells are cheaper to make than multi-Si, due to a great reduction in silicon waste, as this approach does not require sawing from ingots. However, they are also less efficient.

Mono-like-multi Silicon (MLM)

This form was developed in the 2000s and introduced commercially around 2009. Also called cast-mono, this design uses polycrystalline casting chambers with small "seeds" of mono material. The result is a bulk mono-like material that is polycrystalline around the outsides. When sliced for processing, the inner sections are high-efficiency mono-like cells (but square instead of "clipped"), while the outer edges are sold as conventional poly. This production method results in mono-like cells at poly-like prices.

Thin Film

Thin-film technologies reduce the amount of active material in a cell. Most designs sandwich active material between two panes of glass. Since silicon solar panels only use one pane of glass, thin film panels are approximately twice as heavy as crystalline silicon panels, although they have a smaller ecological impact (determined from life cycle analysis).

Cadmium Telluride

Cadmium telluride is the only thin film material so far to rival crystalline silicon in cost/watt. However cadmium is highly toxic and tellurium (anion: "telluride") supplies are limited. The cadmium present in the cells would be toxic if released. However, release is impossible during normal operation of the cells and is unlikely during fires in residential roofs. A square meter of CdTe contains approximately the same amount of Cd as a single C cell nickel-cadmium battery, in a more stable and less soluble form.

Copper Indium Gallium Selenide

Copper indium gallium selenide (CIGS) is a direct band gap material. It has the highest efficiency (~20%) among all commercially significant thin film materials. Traditional methods of fabrication involve vacuum processes including co-evaporation and sputtering. Recent developments at IBM and Nanosolar attempt to lower the cost by using non-vacuum solution processes.

Silicon Thin Film

Silicon thin-film cells are mainly deposited by chemical vapor deposition (typically plasma-enhanced, PE-CVD) from silane gas and hydrogen gas. Depending on the deposition parameters, this can yield amorphous silicon (a-Si or a-Si:H), protocrystalline silicon or nanocrystalline silicon (nc-Si or nc-Si:H), also called microcrystalline silicon.

Amorphous silicon is the most well-developed thin film technology to-date. An amorphous silicon (a-Si) solar cell is made of non-crystalline or microcrystalline silicon. Amorphous silicon has a higher bandgap (1.7 eV) than crystalline silicon (c-Si) (1.1 eV), which means it absorbs the visible part of the solar spectrum more strongly than the higher power density infrared portion of the spectrum. The production of a-Si thin film solar cells uses glass as a substrate and deposits a very thin layer of silicon by plasma-enhanced chemical vapor deposition (PECVD).

Protocrystalline silicon with a low volume fraction of nanocrystalline silicon is optimal for high open circuit voltage. Nc-Si has about the same bandgap as c-Si and nc-Si and a-Si can advantageously be combined in thin layers, creating a layered cell called a tandem cell. The top cell in a-Si absorbs the visible light and leaves the infrared part of the spectrum for the bottom cell in nc-Si.

Gallium Arsenide Thin Film

The semiconductor material Gallium arsenide (GaAs) is also used for single-crystalline thin film solar cells. Although GaAs cells are very expensive, they hold the world's record in efficiency for a single-junction solar cell at 28.8%. GaAs is more commonly used in multijunction photovoltaic cells for concentrated photovoltaics (CPV, HCPV) and for solar panels on spacecrafts, as the industry favours efficiency over cost for space-based solar power. Based on the previous literature and some theoretical analysis, there are several reasons why GaAs has such high power conversion efficiency. First, GaAs bandgap is 1.43ev which is almost ideal for solar cells. Second, because Gallium is a by-product of the smelting of other metals, GaAs cells are relatively insensitive to heat and it can keep high efficiency when temperature is quite high. Third, GaAs has the wide range of design options. Using GaAs as active layer in solar cell, engineers can have multiple choices of other layers which can better generate electrons and holes in GaAs.

Multijunction Cells

Multi-junction cells consist of multiple thin films, each essentially a solar cell grown on top of another, typically using metalorganic vapour phase epitaxy. Each layer has a different band gap energy to allow it to absorb electromagnetic radiation over a different portion of the spectrum. Multi-junction cells were originally developed for special applications such as satellites and space exploration, but are now used increasingly in terrestrial concentrator photovoltaics (CPV), an emerging technology that uses lenses and curved mirrors to concentrate sunlight onto small, highly efficient multi-junction solar cells. By concentrating sunlight up to a thousand times, *High concentrated photovoltaics (HCPV)* has the potential to outcompete conventional solar PV in the future.

Dawn's 10 kW triple-junction gallium arsenide solar array at full extension.

Tandem solar cells based on monolithic, series connected, gallium indium phosphide (GaInP), gallium arsenide (GaAs), and germanium (Ge) p–n junctions, are increasing sales, despite cost pressures. Between December 2006 and December 2007, the cost of 4N gallium metal rose from about $350 per kg to $680 per kg. Additionally, germanium metal prices have risen substantially to $1000–1200 per kg this year. Those materials include gallium (4N, 6N and 7N Ga), arsenic (4N, 6N and 7N) and germanium, pyrolitic boron nitride (pBN) crucibles for growing crystals, and boron oxide, these products are critical to the entire substrate manufacturing industry.

A triple-junction cell, for example, may consist of the semiconductors: GaAs, Ge, and GaInP$_2$. Triple-junction GaAs solar cells were used as the power source of the Dutch four-time World Solar Challenge winners Nuna in 2003, 2005 and 2007 and by the Dutch solar cars Solutra, Twente One and 21Revolution. GaAs based multi-junction devices are the most efficient solar cells to date. On 15 October 2012, triple junction metamorphic cells reached a record high of 44%.

GaInP/Si Dual-junction Solar Cells

In 2016, a new approach was described for producing hybrid photovoltaic wafers combining the high efficiency of III-V multi-junction solar cells with the economies and wealth of experience associated with silicon. The technical complications involved in growing the III-V material on silicon at the required high temperatures, a subject of study for some 30 years, are avoided by epitaxial growth of silicon on GaAs at low temperature by plasma-enhanced chemical vapor deposition (PECVD).

Si single-junction solar cells have been widely studied for decades and are reaching their practical efficiency of ~26% under 1-sun conditions. Increasing this efficiency may require adding more cells with bandgap energy larger than 1.1 eV to the Si cell, allowing to convert short-wavelength photons for generation of additional voltage. A dual-junction solar cell with a band gap of 1.6–1.8 eV as a top cell can reduce thermalization loss, produce a high external radiative efficiency and achieve theoretical efficiencies over 45%. A tandem cell can be fabricated by growing the GaInP and Si cells. Growing them separately can overcome the 4% lattice constant mismatch between Si and the most common III–V layers that prevent direct integration into one cell. The two cells therefore are separated by a transparent glass slide so the lattice mismatch does not cause strain to the system. This creates a cell with four electrical contacts and two junctions that demonstrated an efficiency of 18.1%. With a fill factor (FF) of 76.2%, the Si bottom cell reaches an efficiency of 11.7% (± 0.4) in the tandem device, resulting in a cumulative tandem cell efficiency of 29.8%. This efficiency exceeds the theoretical limit of 29.4% and the record experimental efficiency value of a Si 1-sun solar cell, and is also higher than the record-efficiency 1-sun GaAs device. However, using a GaAs substrate is expensive and not practical. Hence researchers try to make a cell with two electrical contact points and one junction, which does not need a GaAs substrate. This means there will be direct integration of GaInP and Si.

Research in Solar Cells

Perovskite Solar Cells

Perovskite solar cells are solar cells that include a perovskite-structured material as the active layer. Most commonly, this is a solution-processed hybrid organic-inorganic tin or lead halide based material. Efficiencies have increased from below 5% at their first usage in 2009 to over 25% in 2019, making them a very rapidly advancing technology and a hot topic in the solar cell field. Perovskite solar cells are also forecast to be extremely cheap to scale up, making them a very attractive option for commercialisation. So far most types of perovskite solar cells have not reached sufficient operational stability to be commercialised, although many research groups are investigating ways to solve this.

Bifacial Solar Cells

Bifacial solar cell plant in Noto (Senegal), 1988 - Floor painted in white to enhance albedo.

With a transparent rear side, bifacial solar cells can absorb light from both the front and rear sides. Hence, they can produce more electricity than conventional monofacial solar cells. The first patent of bifacial solar cells was filed by Japanese researcher Hiroshi Mori, in 1966. Later, it is said that Russia was the first to deploy bifacial solar cells in their space program in the 1970s. In 1976, the Institute for Solar Energy of the Technical University of Madrid, began a research program for the development of bifacial solar cells led by Prof. Antonio Luque. Based on 1977 US and Spanish patents by Luque, a practical bifacial cell was proposed with a front face as anode and a rear face as cathode; in previously reported proposals and attempts both faces were anodic and interconnection between cells was complicated and expensive. In 1980, Andrés Cuevas, a PhD student in Luque's team, demonstrated experimentally a 50% increase in output power of bifacial solar cells, relative to identically oriented and tilted monofacial ones, when a white background was provided. In 1981 the company Isofoton was founded in Málaga to produce the developed bifacial cells, thus becoming the first industrialization of this PV cell technology. With an initial production capacity of 300 kW/yr. of bifacial solar cells, early landmarks of Isofoton's production were the 20kWp power plant in San Agustín de Guadalix, built in 1986 for Iberdrola, and an off grid installation by 1988 also of 20kWp in the village of Noto Gouye Diama (Senegal) funded by the Spanish international aid and cooperation programs.

Due to the reduced manufacturing cost, companies have again started to produce commercial bifacial modules since 2010. By 2017, there were at least eight certified PV manufacturers providing bifacial modules in North America. It has been predicted by the International Technology Roadmap for Photovoltaics (ITRPV) that the global market share of bifacial technology will expand from less than 5% in 2016 to 30% in 2027.

Due to the significant interest in the bifacial technology, a recent study has investigated the performance and optimization of bifacial solar modules worldwide. The results indicate that, across the globe, ground-mounted bifacial modules can only offer ~10% gain in annual electricity yields compared to the monofacial counterparts for a ground albedo coefficient of 25% (typical for concrete and vegetation groundcovers). However, the gain can be increased to ~30% by elevating the module 1 m above the ground and enhancing the ground albedo coefficient to 50%. Sun *et al.* also derived a set of empirical equations that can optimize bifacial solar modules analytically.

An online simulation tool is available to model the performance of bifacial modules in any arbitrary location across the entire world. It can also optimize bifacial modules as a function of tilt angle, azimuth angle, and elevation above the ground.

Intermediate Band

Intermediate band photovoltaics in solar cell research provides methods for exceeding the Shockley–Queisser limit on the efficiency of a cell. It introduces an intermediate band (IB) energy level in between the valence and conduction bands. Theoretically, introducing an IB allows two photons with energy less than the bandgap to excite an electron from the valence band to the conduction band. This increases the induced photocurrent and thereby efficiency.

Luque and Marti first derived a theoretical limit for an IB device with one midgap energy level using detailed balance. They assumed no carriers were collected at the IB and that the device was under full concentration. They found the maximum efficiency to be 63.2%, for a bandgap of 1.95eV with the IB 0.71eV from either the valence or conduction band. Under one sun illumination the limiting efficiency is 47%.

Liquid Inks

In 2014, researchers at California NanoSystems Institute discovered using kesterite and perovskite improved electric power conversion efficiency for solar cells.

Upconversion and Downconversion

Photon upconversion is the process of using two low-energy (*e.g.*, infrared) photons to produce one higher energy photon; downconversion is the process of using one high energy photon (*e.g.,*, ultraviolet) to produce two lower energy photons. Either of these techniques could be used to produce higher efficiency solar cells by allowing solar photons to be more efficiently used. The difficulty, however, is that the conversion efficiency of existing phosphors exhibiting up- or down-conversion is low, and is typically narrow band.

One upconversion technique is to incorporate lanthanide-doped materials (Er^{3+}, Yb^{3+}, Ho^{3+} or a combination), taking advantage of their luminescence to convert infrared radiation to visible light. Upconversion process occurs when two infrared photons are absorbed by rare-earth ions to generate a (high-energy) absorbable photon. As example, the energy transfer upconversion process (ETU), consists in successive transfer processes between excited ions in the near infrared. The upconverter material could be placed below the solar cell to absorb the infrared light that passes through the silicon. Useful ions are most commonly found in the trivalent state. Er^+ ions have been the most used. Er^{3+} ions absorb solar radiation around 1.54 μm. Two Er^{3+} ions that have absorbed this radiation can interact with each other through an upconversion process. The excited ion emits light above the Si bandgap that is absorbed by the solar cell and creates an additional electron–hole pair that can generate current. However, the increased efficiency was small. In addition, fluoroindate glasses have low phonon energy and have been proposed as suitable matrix doped with Ho^{3+} ions.

Light-absorbing Dyes

Dye-sensitized solar cells (DSSCs) are made of low-cost materials and do not need elaborate manufacturing equipment, so they can be made in a DIY fashion. In bulk it should be significantly less expensive than older solid-state cell designs. DSSC's can be engineered into flexible sheets and although its conversion efficiency is less than the best thin film cells, its price/performance ratio may be high enough to allow them to compete with fossil fuel electrical generation.

Typically a ruthenium metalorganic dye (Ru-centered) is used as a monolayer of light-absorbing material. The dye-sensitized solar cell depends on a mesoporous layer of nanoparticulate titanium dioxide to greatly amplify the surface area (200–300 m^2/g TiO_2, as compared to approximately 10 m^2/g of flat single crystal). The photogenerated electrons from the light absorbing dye are passed on to the n-type TiO_2 and the holes are absorbed by an electrolyte on the other side of the dye. The circuit is completed by a redox couple in the electrolyte, which can be liquid or solid. This type of cell allows more flexible use of materials and is typically manufactured by screen printing or ultrasonic nozzles, with the potential for lower processing costs than those used for bulk solar cells. However, the dyes in these cells also suffer from degradation under heat and UV light and the cell casing is difficult to seal due to the solvents used in assembly. The first commercial shipment of DSSC solar modules occurred in July 2009 from G24i Innovations.

Quantum Dots

Quantum dot solar cells (QDSCs) are based on the Gratzel cell, or dye-sensitized solar cell architecture, but employ low band gap semiconductor nanoparticles, fabricated with crystallite sizes small enough to form quantum dots (such as CdS, CdSe, Sb_2S_3, PbS, etc.), instead of organic or organo-metallic dyes as light absorbers. Due to the toxicity associated with Cd and Pb based compounds there are also a series of "green" QD sensitizing materials in development (such as $CuInS_2$, $CuInSe_2$ and CuInSeS). QD's size quantization allows for the band gap to be tuned by simply changing particle size. They also have high extinction coefficients and have shown the possibility of multiple exciton generation.

In a QDSC, a mesoporous layer of titanium dioxide nanoparticles forms the backbone of the cell, much like in a DSSC. This TiO_2 layer can then be made photoactive by coating with semiconductor quantum dots using chemical bath deposition, electrophoretic deposition or successive ionic layer adsorption and reaction. The electrical circuit is then completed through the use of a liquid or solid redox couple. The efficiency of QDSCs has increased to over 5% shown for both liquid-junction and solid state cells, with a reported peak efficiency of 11.91%. In an effort to decrease production costs, the Prashant Kamat research group demonstrated a solar paint made with TiO_2 and CdSe that can be applied using a one-step method to any conductive surface with efficiencies over 1%. However, the absorption of quantum dots (QDs) in QDSCs is weak at room temperature. The plasmonic nanoparticles can be utilized to address the weak absorption of QDs (e.g., nanostars). Adding an external infrared pumping sources to excite intraband and interband transition of QDs is another solution.

Organic/Polymer Solar Cells

Organic solar cells and polymer solar cells are built from thin films (typically 100 nm) of organic semiconductors including polymers, such as polyphenylene vinylene and small-molecule compounds like copper phthalocyanine (a blue or green organic pigment) and carbon fullerenes and fullerene derivatives such as PCBM.

They can be processed from liquid solution, offering the possibility of a simple roll-to-roll printing process, potentially leading to inexpensive, large-scale production. In addition, these cells could be beneficial for some applications where mechanical flexibility and disposability are important. Current cell efficiencies are, however, very low, and practical devices are essentially non-existent.

Energy conversion efficiencies achieved to date using conductive polymers are very low compared to inorganic materials. However, Konarka Power Plastic reached efficiency of 8.3% and organic tandem cells in 2012 reached 11.1%.

The active region of an organic device consists of two materials, one electron donor and one electron acceptor. When a photon is converted into an electron hole pair, typically in the donor material, the charges tend to remain bound in the form of an exciton, separating when the exciton diffuses to the donor-acceptor interface, unlike most other solar cell types. The short exciton diffusion lengths of most polymer systems tend to limit the efficiency of such devices. Nanostructured interfaces, sometimes in the form of bulk heterojunctions, can improve performance.

In 2011, MIT and Michigan State researchers developed solar cells with a power efficiency close to 2% with a transparency to the human eye greater than 65%, achieved by selectively absorbing the ultraviolet and near-infrared parts of the spectrum with small-molecule compounds. Researchers at UCLA more recently developed an analogous polymer solar cell, following the same approach, that is 70% transparent and has a 4% power conversion efficiency. These lightweight, flexible cells can be produced in bulk at a low cost and could be used to create power generating windows.

In 2013, researchers announced polymer cells with some 3% efficiency. They used block copolymers, self-assembling organic materials that arrange themselves into distinct layers. The research focused on P3HT-b-PFTBT that separates into bands some 16 nanometers wide.

Adaptive Cells

Adaptive cells change their absorption/reflection characteristics depending to respond to environmental conditions. An adaptive material responds to the intensity and angle of incident light. At the part of the cell where the light is most intense, the cell surface changes from reflective to adaptive, allowing the light to penetrate the cell. The other parts of the cell remain reflective increasing the retention of the absorbed light within the cell.

In 2014, a system was developed that combined an adaptive surface with a glass substrate that redirect the absorbed to a light absorber on the edges of the sheet. The system also includes an array of fixed lenses/mirrors to concentrate light onto the adaptive surface. As the day continues, the concentrated light moves along the surface of the cell. That surface switches from reflective to adaptive when the light is most concentrated and back to reflective after the light moves along.

Surface Texturing

Solar Impulse aircraft are Swiss-designed single-seat
monoplanes powered entirely from photovoltaic cells.

For the past years, researchers have been trying to reduce the price of solar cells while maximizing efficiency. Thin-film solar cell is a cost-effective second generation solar cell with much reduced thickness at the expense of light absorption efficiency. Efforts to maximize light absorption efficiency with reduced thickness have been made. Surface texturing is one of techniques used to reduce optical losses to maximize light absorbed. Currently, surface texturing techniques on silicon photovoltaics are drawing much attention. Surface texturing could be done in multiple ways. Etching single crystalline silicon substrate can produce randomly distributed square based pyramids on the surface using anisotropic etchants. Recent studies show that c-Si wafers could be etched down to form nano-scale inverted pyramids. Multicrystalline silicon solar cells, due to poorer crystallographic quality, are less effective than single crystal solar cells, but mc-Si solar cells are still being used widely due to less manufacturing difficulties. It is reported that multicrystalline solar cells can be surface-textured to yield solar energy conversion efficiency comparable to that of monocrystalline silicon cells, through isotropic etching or photolithography techniques. Incident light rays onto a textured surface do not reflect back out to the air as opposed to rays onto a flat surface. Rather some light rays are bounced back onto the other surface again due to the geometry of the surface. This process significantly improves light to electricity conversion efficiency, due to increased light absorption. This texture effect as well as the interaction with other interfaces in the PV module is a challenging optical simulation task. A particularly efficient method for modeling and optimization is the OPTOS formalism. In 2012, researchers at MIT reported that c-Si films textured with nanoscale inverted pyramids could achieve light absorption comparable to 30 times thicker planar c-Si. In combination with anti-reflective coating, surface texturing technique can effectively trap light rays within a thin film silicon solar cell. Consequently, required thickness for solar cells decreases with the increased absorption of light rays.

Encapsulation

Solar cells are commonly encapsulated in a transparent polymeric resin to protect the delicate solar cell regions for coming into contact with moisture, dirt, ice, and other conditions expected either during operation or when used outdoors. The encapsulants are commonly made from polyvinyl acetate or glass. Most encapsulants are uniform in structure and composition, which increases light collection owing to light trapping from total internal reflection of light within the resin. Research has been conducted into structuring the encapsulant to provide further collection of light. Such encapsulants have included roughened glass surfaces, diffractive elements, prism arrays, air prisms, v-grooves, diffuse elements, as well as multi-directional waveguide arrays. Prism arrays show an overall 5% increase in the total solar energy conversion. Arrays of vertically aligned broadband waveguides provide a 10% increase at normal incidence, as well as wide-angle collection enhancement of up to 4%, with optimized structures yielding up to a 20% increase in short circuit current. Active coatings that convert infrared light into visible light have shown a 30% increase. Nanoparticle coatings inducing plasmonic light scattering increase wide-angle conversion efficiency up to 3%. Optical structures have also been created in encapsulation materials to effectively "cloak" the metallic front contacts.

Manufacture

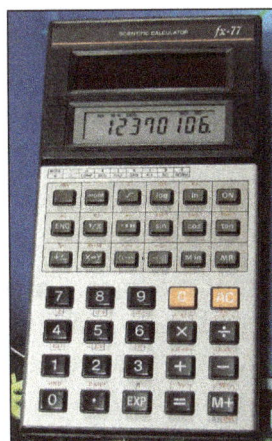

Early solar-powered calculator.

Solar cells share some of the same processing and manufacturing techniques as other semiconductor devices. However, the stringent requirements for cleanliness and quality control of semiconductor fabrication are more relaxed for solar cells, lowering costs.

Polycrystalline silicon wafers are made by wire-sawing block-cast silicon ingots into 180 to 350 micrometer wafers. The wafers are usually lightly p-type-doped. A surface diffusion of n-type dopants is performed on the front side of the wafer. This forms a p–n junction a few hundred nanometers below the surface.

Anti-reflection coatings are then typically applied to increase the amount of light coupled into the solar cell. Silicon nitride has gradually replaced titanium dioxide as the preferred material, because of its excellent surface passivation qualities. It prevents carrier recombination at the cell surface. A layer several hundred nanometers thick is applied using PECVD. Some solar cells have textured front surfaces that, like anti-reflection coatings, increase the amount of light reaching the wafer. Such surfaces were first applied to single-crystal silicon, followed by multicrystalline silicon somewhat later.

A full area metal contact is made on the back surface, and a grid-like metal contact made up of fine "fingers" and larger "bus bars" are screen-printed onto the front surface using a silver paste. This is an evolution of the so-called "wet" process for applying electrodes, first described in a US patent filed in 1981 by Bayer AG. The rear contact is formed by screen-printing a metal paste, typically aluminium. Usually this contact covers the entire rear, though some designs employ a grid pattern. The paste is then fired at several hundred degrees Celsius to form metal electrodes in ohmic contact with the silicon. Some companies use an additional electro-plating step to increase efficiency. After the metal contacts are made, the solar cells are interconnected by flat wires or metal ribbons, and assembled into modules or "solar panels". Solar panels have a sheet of tempered glass on the front, and a polymer encapsulation on the back.

Disposal

Solar cells degrade over time and lose their efficiency. Solar cells in extreme climates, such as desert or polar, are more prone to degradation due to exposure to harsh UV light and snow loads respectively. Usually, solar panels are given a lifespan of 25–30 years before they get decommissioned.

The International Renewable Energy Agency estimated that the amount of solar panel waste generated in 2016 was 43,500–250,000 metric tons. This number is estimated to increase substantially by 2030, reaching an estimated waste volume of 60–78 million metric tons in 2050.

Recycling

Solar panels are recycled through different methods. The recycling process include a three step process, module recycling, cell recycling and waste handling, to break down Si modules and recover various materials. The recovered metals and Si are re-usable to the solar industry and generate $11–12.10/module in revenue at today's prices for Ag and solar-grade Si.

Some solar modules contains toxic materials like lead and cadmium which, when broken, could possible leach into the soil and contaminate the environment. The First Solar panel recycling plant opened in Rousset, France in 2018. It was set to recycle 1300 tonnes of solar panel waste a year, and can increase its capacity to 4000 tonnes.

Applications

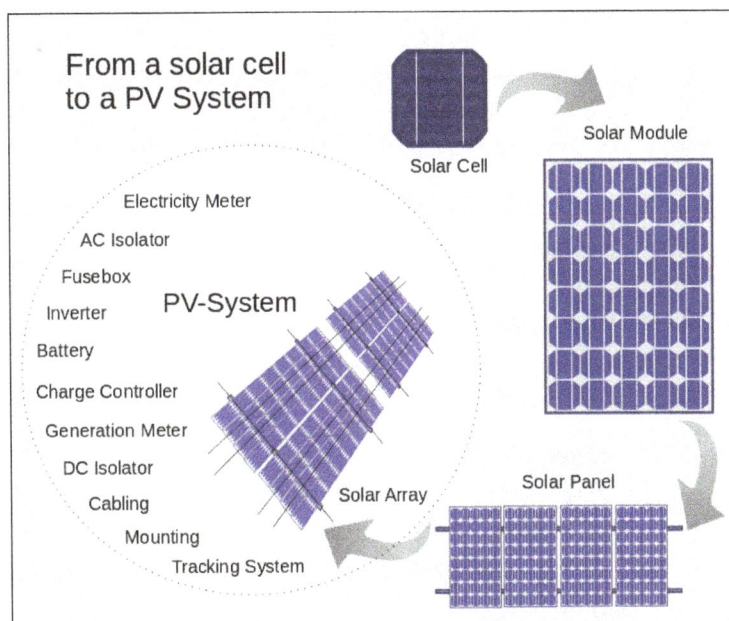

From a solar cell to a PV system. Diagram of the possible components of a photovoltaic system.

Assemblies of solar cells are used to make solar modules that generate electrical power from sunlight, as distinguished from a "solar thermal module" or "solar hot water panel". A solar array generates solar power using solar energy.

Cells, Modules, Panels and Systems

Multiple solar cells in an integrated group, all oriented in one plane, constitute a solar photovoltaic panel or module. Photovoltaic modules often have a sheet of glass on the sun-facing side, allowing light to pass while protecting the semiconductor wafers. Solar cells are usually connected in series and parallel circuits or series in modules, creating an additive voltage. Connecting cells in parallel yields a higher current; however, problems such as shadow effects can shut down the weaker (less illuminated) parallel string (a number of series connected cells) causing substantial power loss and possible damage because of the reverse bias applied to the shadowed cells by their illuminated partners. Strings of series cells are usually handled independently and not connected in parallel, though as of 2014, individual power boxes are often supplied for each module, and are connected in parallel. Although modules can be interconnected to create an array with the desired peak DC voltage and loading current capacity, using independent MPPTs (maximum power point trackers) is preferable. Otherwise, shunt diodes can reduce shadowing power loss in arrays with series/parallel connected cells.

Typical PV system prices in 2013 in selected countries ($/W)								
USD/W	Australia	China	France	Germany	Italy	Japan	United Kingdom	United States
Residential	1.8	1.5	4.1	2.4	2.8	4.2	2.8	4.9
Commercial	1.7	1.4	2.7	1.8	1.9	3.6	2.4	4.5
Utility-scale	2.0	1.4	2.2	1.4	1.5	2.9	1.9	3.3

Thin-film Solar Cell

A thin-film solar cell is a second generation solar cell that is made by depositing one or more thin layers, or thin film (TF) of photovoltaic material on a substrate, such as glass, plastic or metal. Thin-film solar cells are commercially used in several technologies, including cadmium telluride (CdTe), copper indium gallium diselenide (CIGS), and amorphous thin-film silicon (a-Si, TF-Si).

Film thickness varies from a few nanometers (nm) to tens of micrometers (µm), much thinner than thin-film's rival technology, the conventional, first-generation crystalline silicon solar cell (c-Si), that uses wafers of up to 200 µm thick. This allows thin film cells to be flexible, and lower in weight. It is used in building integrated photovoltaics and as semi-transparent, photovoltaic glazing material that can be laminated onto windows. Other commercial applications use rigid thin film solar panels (interleaved between two panes of glass) in some of the world's largest photovoltaic power stations.

Thin-film technology has always been cheaper but less efficient than conventional c-Si technology. However, it has significantly improved over the years. The lab cell efficiency for CdTe and CIGS is now beyond 21 percent, outperforming multicrystalline silicon, the dominant material currently used in most solar PV systems. Accelerated life testing of thin film modules under laboratory conditions measured a somewhat faster degradation compared to conventional PV, while a lifetime of 20 years or more is generally expected. Despite these enhancements, market-share of thin-film never reached more than 20 percent in the last two decades and has been declining in recent years to about 9 percent of worldwide photovoltaic installations in 2013.

Other thin-film technologies that are still in an early stage of ongoing research or with limited commercial availability are often classified as emerging or third generation photovoltaic cells and include organic, and dye-sensitized, as well as quantum dot, copper zinc tin sulfide, nanocrystal, micromorph, and perovskite solar cells.

Materials

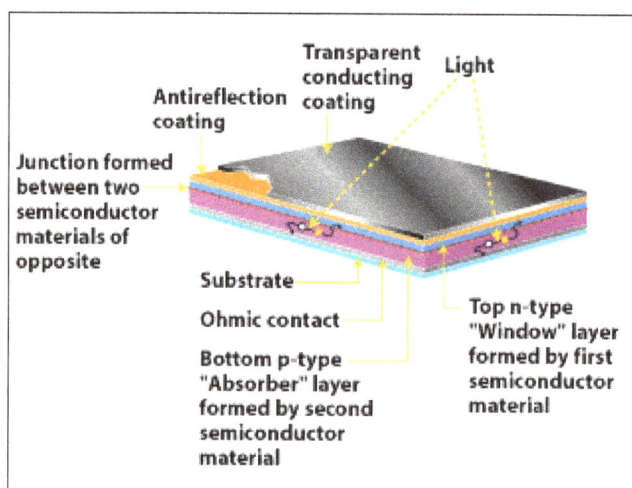

Cross-section of a TF cell.

Thin-film technologies reduce the amount of active material in a cell. Most sandwich active material between two panes of glass. Since silicon solar panels only use one pane of glass, thin film panels are approximately twice as heavy as crystalline silicon panels, although they have a smaller ecological impact (determined from life cycle analysis). The majority of film panels have 2-3 percentage points lower conversion efficiencies than crystalline silicon. Cadmium telluride (CdTe), copper indium gallium selenide (CIGS) and amorphous silicon (a-Si) are three thin-film technologies often used for outdoor applications.

Cadmium Telluride

Cadmium telluride (CdTe) is the predominant thin film technology. With about 5 percent of worldwide PV production, it accounts for more than half of the thin film market. The cell's lab efficiency has also increased significantly in recent years and is on a par with CIGS thin film and close to the efficiency of multi-crystalline silicon as of 2013. Also, CdTe has the lowest Energy payback time of all mass-produced PV technologies, and can be as short as eight months in favorable locations. A prominent manufacturer is the US-company First Solar based in Tempe, Arizona, that produces CdTe-panels with an efficiency of about 14 percent at a reported cost of $0.59 per watt.

Although the toxicity of cadmium may not be that much of an issue and environmental concerns completely resolved with the recycling of CdTe modules at the end of their life time, there are still uncertainties and the public opinion is skeptical towards this technology. The usage of rare materials may also become a limiting factor to the industrial scalability of CdTe thin film technology. The rarity of tellurium—of which telluride is the anionic form—is comparable to that of platinum in the earth's crust and contributes significantly to the module's cost.

Copper Indium Gallium Selenide

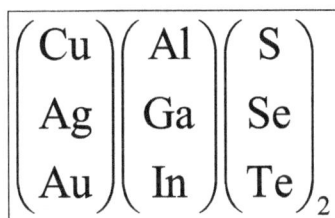

$$\begin{pmatrix} Cu \\ Ag \\ Au \end{pmatrix} \begin{pmatrix} Al \\ Ga \\ In \end{pmatrix} \begin{pmatrix} S \\ Se \\ Te \end{pmatrix}_2$$

Possible combinations of Group-(XI, XIII, XVI) elements in the periodic table that yield a compound showing photovoltaic effect: *Cu, Ag, Au – Al, Ga, In – S, Se, Te.*

A copper indium gallium selenide solar cell or CIGS cell uses an absorber made of copper, indium, gallium, selenide (CIGS), while gallium-free variants of the semiconductor material are abbreviated CIS. It is one of three mainstream thin-film technologies, the other two being cadmium telluride and amorphous silicon, with a lab-efficiency above 20 percent and a share of 2 percent in the overall PV market in 2013. A prominent manufacturer of cylindrical CIGS-panels was the now-bankrupt company Solyndra in Fremont, California. Traditional methods of fabrication involve vacuum processes including co-evaporation and sputtering. In 2008, IBM and Tokyo Ohka Kogyo Co., Ltd. (TOK) announced they had developed a new, non-vacuum, solution-based manufacturing process for CIGS cells and are aiming for efficiencies of 15% and beyond.

Hyperspectral imaging has been used to characterize these cells. Researchers from IRDEP (Institute of Research and Development in Photovoltaic Energy) in collaboration with Photon etc., were able to determine the splitting of the quasi-Fermi level with photoluminescence mapping while the electroluminescence data were used to derive the external quantum efficiency (EQE). Also, through a light beam induced current (LBIC) cartography experiment, the EQE of a microcrystalline CIGS solar cell could be determined at any point in the field of view.

As of April 2019, current conversion efficiency record for a laboratory CIGS cell stands at 22.9%.

Silicon

Three major silicon-based module designs dominate:

- Amorphous silicon cells.

- Amorphous/microcrystalline tandem cells (micromorph).

- Thin-film polycrystalline silicon on glass.

Amorphous Silicon

Amorphous silicon (a-Si) is a non-crystalline, allotropic form of silicon and the most well-developed thin film technology to-date. Thin-film silicon is an alternative to conventional *wafer* (or *bulk*) crystalline silicon. While chalcogenide-based CdTe and CIS thin films cells have been developed in the lab with great success, there is still industry interest in silicon-based thin film cells. Silicon-based devices exhibit fewer problems than their CdTe and CIS counterparts such as toxicity and humidity issues with CdTe cells and low manufacturing yields of CIS due to material complexity. Additionally, due to political resistance to the use non- "green" materials in solar energy production, there is no stigma in the use of standard silicon.

This type of thin-film cell is mostly fabricated by a technique called plasma-enhanced chemical vapor deposition. It uses a gaseous mixture of silane (SiH_4) and hydrogen to deposit a very thin layer of only 1 micrometre (μm) of silicon on a substrate, such as glass, plastic or metal, that has already been coated with a layer of transparent conducting oxide. Other methods used to deposit amorphous silicon on a substrate include sputtering and hot wire chemical vapor deposition techniques.

a-Si is attractive as a solar cell material because it's an abundant, non-toxic material. It requires a low processing temperature and enables a scalable production upon a flexible, low-cost substrate with little silicon material required. Due to its bandgap of 1.7 eV, amorphous silicon also absorbs a very broad range of the light spectrum, that includes infrared and even some ultraviolet and performs very well at weak light. This allows the cell to generate power in the early morning, or late afternoon and on cloudy and rainy days, contrary to crystalline silicon cells, that are significantly less efficient when exposed at diffuse and indirect daylight.

However, the efficiency of an a-Si cell suffers a significant drop of about 10 to 30 percent during the first six months of operation. This is called the Staebler-Wronski effect (SWE) – a typical loss in electrical output due to changes in photoconductivity and dark conductivity caused by prolonged exposure to sunlight. Although this degradation is perfectly reversible upon annealing at or above 150 °C, conventional c-Si solar cells do not exhibit this effect in the first place.

Its basic electronic structure is the p-i-n junction. The amorphous structure of a-Si implies high inherent disorder and dangling bonds, making it a bad conductor for charge carriers. These dangling bonds act as recombination centers that severely reduce carrier lifetime. A p-i-n structure is usually used, as opposed to an n-i-p structure. This is because the mobility of electrons in a-Si:H is roughly 1 or 2 orders of magnitude larger than that of holes, and thus the collection rate of electrons moving from the n- to p-type contact is better than holes moving from p- to n-type contact. Therefore, the p-type layer should be placed at the top where the light intensity is stronger, so that the majority of the charge carriers crossing the junction are electrons.

Tandem-cell using a-Si/μc-Si

A layer of amorphous silicon can be combined with layers of other allotropic forms of silicon to produce a multi-junction solar cell. When only two layers (two p-n junctions) are combined, it is called a *tandem-cell*. By stacking these layers on top of one other, a broader range of the light spectra is absorbed, improving the cell's overall efficiency.

In micromorphous silicon, a layer of microcrystalline silicon (μc-Si) is combined with amorphous silicon, creating a tandem cell. The top a-Si layer absorbs the visible light, leaving the infrared part to the bottom μc-Si layer. The micromorph stacked-cell concept was pioneered and patented at the Institute of Microtechnology (IMT) of the Neuchâtel University in Switzerland, and was licensed to TEL Solar. A new world record PV module based on the micromorph concept with 12.24% module efficiency was independently certified in July 2014.

Because all layers are made of silicon, they can be manufactured using PECVD. The band gap of a-Si is 1.7 eV and that of c-Si is 1.1 eV. The c-Si layer can absorb red and infrared light. The best efficiency can be achieved at transition between a-Si and c-Si. As nanocrystalline silicon (nc-Si) has about the same bandgap as c-Si, nc-Si can replace c-Si.

Tandem-cell using a-Si/pc-Si

Amorphous silicon can also be combined with protocrystalline silicon (pc-Si) into a tandem-cell. Protocrystalline silicon with a low volume fraction of nanocrystalline silicon is optimal for high open-circuit voltage. These types of silicon present dangling and twisted bonds, which results in deep defects (energy levels in the bandgap) as well as deformation of the valence and conduction bands (band tails).

Polycrystalline Silicon on Glass

A new attempt to fuse the advantages of bulk silicon with those of thin-film devices is thin film polycrystalline silicon on glass. These modules are produced by depositing an antireflection coating and doped silicon onto textured glass substrates using plasma-enhanced chemical vapor deposition (PECVD). The texture in the glass enhances the efficiency of the cell by approximately 3% by reducing the amount of incident light reflecting from the solar cell and trapping light inside the solar cell. The silicon film is crystallized by an annealing step, temperatures of 400–600 Celsius, resulting in polycrystalline silicon.

These new devices show energy conversion efficiencies of 8% and high manufacturing yields of >90%. Crystalline silicon on glass (CSG), where the polycrystalline silicon is 1–2 micrometres, is noted for its stability and durability; the use of thin film techniques also contributes to a cost savings over bulk photovoltaics. These modules do not require the presence of a transparent conducting oxide layer. This simplifies the production process twofold; not only can this step be skipped, but the absence of this layer makes the process of constructing a contact scheme much simpler. Both of these simplifications further reduce the cost of production. Despite the numerous advantages over alternative design, production cost estimations on a per unit area basis show that these devices are comparable in cost to single-junction amorphous thin film cells.

Gallium Arsenide

The semiconductor material gallium arsenide (GaAs) is also used for single-crystalline thin film solar cells. Although GaAs cells are very expensive, they hold the world record for the highest-efficiency, single-junction solar cell at 28.8%. GaAs is more commonly used in multi-junction solar cells for solar panels on spacecrafts, as the larger power to weight ratio lowers the launch costs in space-based solar power (InGaP/(In)GaAs/Ge cells). They are also used in concentrator photovoltaics, an emerging technology best suited for locations that receive much sunlight, using lenses to focus sunlight on a much smaller, thus less expensive GaAs concentrator solar cell.

Emerging Photovoltaics

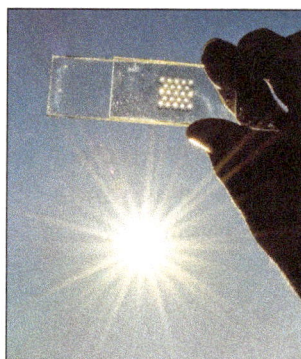

An experimental silicon based solar cell developed at the Sandia National Laboratories.

The National Renewable Energy Laboratory (NREL) classifies a number of thin-film technologies as emerging photovoltaics—most of them have not yet been commercially applied and are still in the research or development phase. Many use organic materials, often organometallic compounds as well as inorganic substances. Despite the fact that their efficiencies had been low and the stability of the absorber material was often too short for commercial applications, there is a lot of research invested into these technologies as they promise to achieve the goal of producing low-cost, high-efficient solar cells.

Emerging photovoltaics, often called third generation photovoltaic cells, include:

- Copper zinc tin sulfide solar cell (CZTS), and derivates CZTSe and CZTSSe.
- Dye-sensitized solar cell, also known as "Grätzel cell".
- Organic solar cell.
- Perovskite solar cell.
- Quantum dot solar cell.

Especially the achievements in the research of perovskite cells have received tremendous attention in the public, as their research efficiencies recently soared above 20 percent. They also offer a wide spectrum of low-cost applications. In addition, another emerging technology, concentrator photovoltaics (CPV), uses high-efficient, multi-junction solar cells in combination with optical lenses and a tracking system.

Efficiencies

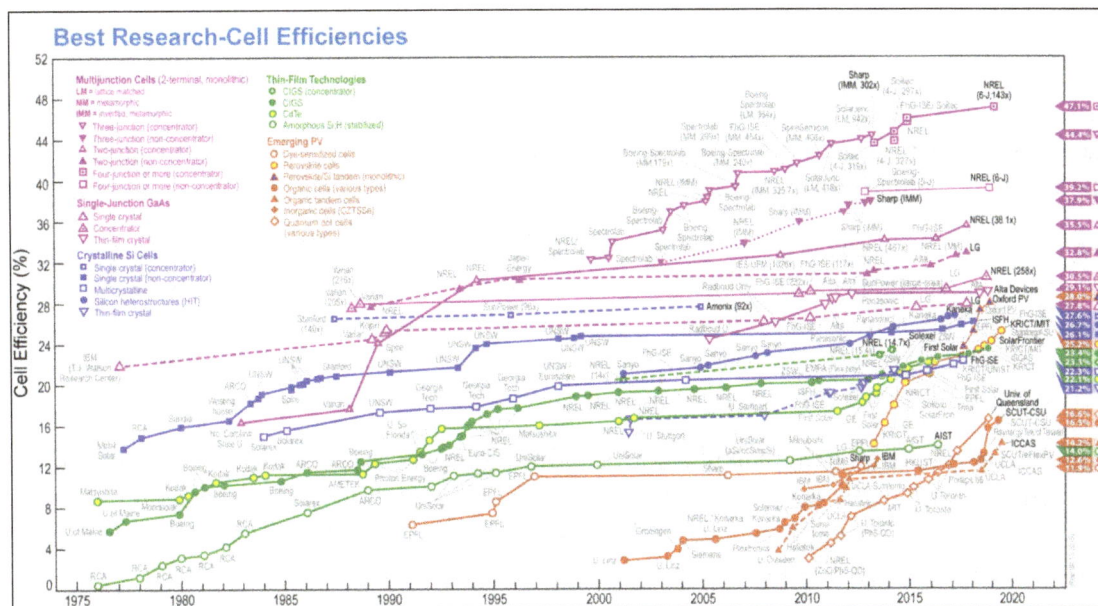

Solar cell efficiencies of various cell technologies as tracked by NREL.

Incremental improvements in efficiency began with the invention of the first modern silicon solar cell in 1954. By 2010 these steady improvements had resulted in modules capable of converting 12 to 18 percent of solar radiation into electricity. The improvements to efficiency have continued to accelerate in the years since 2010, as shown in the accompanying chart.

Cells made from newer materials tend to be less efficient than bulk silicon, but are less expensive to produce. Their quantum efficiency is also lower due to reduced number of collected charge carriers per incident photon.

The performance and potential of thin-film materials are high, reaching cell efficiencies of 12–20%; prototype module efficiencies of 7–13%; and production modules in the range of 9%. The thin film cell prototype with the best efficiency yields 20.4% (First Solar), comparable to the best conventional solar cell prototype efficiency of 25.6% from Panasonic.

NREL once predicted that costs would drop below $100/m² in volume production, and could later fall below $50/m².

A new record for thin film solar cell efficiency of 22.3% has been achieved by solar frontier the world's largest cis solar energy provider. In joint research with the New Energy and Industrial Technology Development Organization (NEDO) of Japan, Solar Frontier achieved 22.3% conversion efficiency on a 0.5 cm² cell using its CIS technology. This is an increase of 0.6 percentage points over the industry's previous thin-film record of 21.7%.

Absorption

Multiple techniques have been employed to increase the amount of light that enters the cell and reduce the amount that escapes without absorption. The most obvious technique is to minimizing the top contact coverage of the cell surface, reducing the area that blocks light from reaching the cell.

The weakly absorbed long wavelength light can be obliquely coupled into silicon and traverses the film several times to enhance absorption.

Multiple methods have been developed to increase absorption by reducing the number of incident photons being reflected away from the cell surface. An additional anti-reflective coating can cause destructive interference within the cell by modulating the refractive index of the surface coating. Destructive interference eliminates the reflective wave, causing all incident light to enter the cell.

Surface texturing is another option for increasing absorption, but increases costs. By applying a texture to the active material's surface, the reflected light can be refracted into striking the surface again, thus reducing reflectance. For example, black silicon texturing by reactive ion etching(RIE) is an effective and economic approach to increase the absorption of thin-film silicon solar cells. A textured backreflector can prevent light from escaping through the rear of the cell.

In addition to surface texturing, the plasmonic light-trapping scheme attracted a lot of attention to aid photocurrent enhancement in thin film solar cells. This method makes use of collective oscillation of excited free electrons in noble metal nanoparticles, which are influenced by particle shape, size and dielectric properties of the surrounding medium.

In addition to minimizing reflective loss, the solar cell material itself can be optimized to have higher chance of absorbing a photon that reaches it. Thermal processing techniques can significantly enhance the crystal quality of silicon cells and thereby increase efficiency. Layering thin-film cells to create a multi-junction solar cell can also be done. Each layer's band gap can be designed to best absorb a different range of wavelengths, such that together they can absorb a greater spectrum of light.

Further advancement into geometric considerations can exploit nanomaterial dimensionality. Large, parallel nanowire arrays enable long absorption lengths along the length of the wire while maintaining short minority carrier diffusion lengths along the radial direction. Adding nanoparticles between the nanowires allows conduction. The natural geometry of these arrays forms a textured surface that traps more light.

Organic Solar Cell

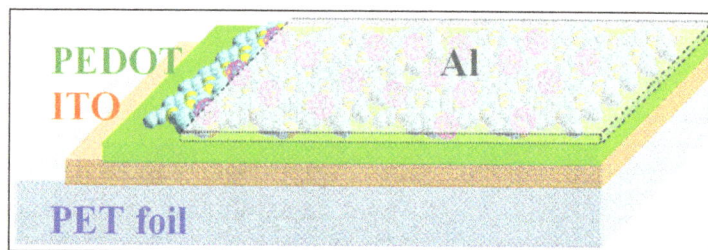

Schematic of plastic solar cells. PET – polyethylene terephthalate, ITO – indium tin oxide, PEDOT:PSS – poly (3,4-ethylenedioxythiophene), active layer (usually a polymer:fullerene blend), Al – aluminium.

An organic solar cell (OSC) or plastic solar cell is a type of photovoltaic that uses organic electronics, a branch of electronics that deals with conductive organic polymers or small organic molecules, for light absorption and charge transport to produce electricity from sunlight by the photovoltaic effect. Most organic photovoltaic cells are polymer solar cells.

Organic Photovoltaic manufactured by the company, Solarmer.

The molecules used in organic solar cells are solution-processable at high throughput and are cheap, resulting in low production costs to fabricate a large volume. Combined with the flexibility of organic molecules, organic solar cells are potentially cost-effective for photovoltaic applications. Molecular engineering (e.g. changing the length and functional group of polymers) can change the band gap, allowing for electronic tunability. The optical absorption coefficient of organic molecules is high, so a large amount of light can be absorbed with a small amount of materials, usually on the order of hundreds of nanometers. The main disadvantages associated with organic photovoltaic cells are low efficiency, low stability and low strength compared to inorganic photovoltaic cells such as silicon solar cells.

Compared to silicon-based devices, polymer solar cells are lightweight (which is important for small autonomous sensors), potentially disposable and inexpensive to fabricate (sometimes using printed electronics), flexible, customizable on the molecular level and potentially have less adverse environmental impact. Polymer solar cells also have the potential to exhibit transparency, suggesting applications in windows, walls, flexible electronics, etc. An example device is shown in figure The disadvantages of polymer solar cells are also serious: they offer about 1/3 of the efficiency of hard materials, and experience substantial photochemical degradation.

Polymer solar cells inefficiency and stability problems, combined with their promise of low costs and increased efficiency made them a popular field in solar cell research. As of 2015, polymer solar cells were able to achieve over 10% efficiency via a tandem structure. In 2018, a record breaking efficiency for organic photovoltaics of 17.3% was reached via tandem structure.

A photovoltaic cell is a specialized semiconductor diode that converts light into direct current (DC) electricity. Depending on the band gap of the light-absorbing material, photovoltaic cells can also convert low-energy, infrared (IR) or high-energy, ultraviolet (UV) photons into DC electricity. A common characteristic of both the small molecules and polymers used as the light-absorbing material in photovoltaics is that they all have large conjugated systems. A conjugated system is formed where carbon atoms covalently bond with alternating single and double bonds. These hydrocarbons' electrons pz orbitals delocalize and form a delocalized bonding π orbital with a π^* antibonding orbital. The delocalized π orbital is the highest occupied molecular orbital (HOMO), and the π^* orbital is the lowest unoccupied molecular orbital (LUMO). In organic semiconductor physics, the HOMO takes the role of the valence band while the LUMO serves as the conduction band. The energy separation between the HOMO and LUMO energy levels is considered the band gap of organic electronic materials and is typically in the range of 1–4 eV.

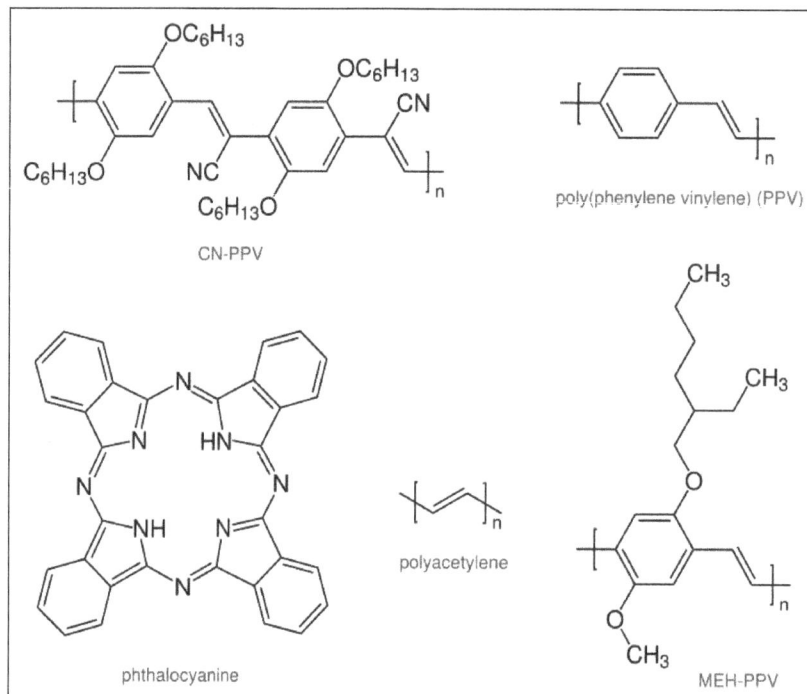

Examples of organic photovoltaic materials.

All light with energy greater than the band gap of the material can be absorbed, though there is a trade-off to reducing the band gap as photons absorbed with energies higher than the band gap will thermally give off its excess energy, resulting in lower voltages and power conversion efficiencies. When these materials absorb a photon, an excited state is created and confined to a molecule or a region of a polymer chain. The excited state can be regarded as an exciton, or an electron-hole pair bound together by electrostatic interactions. In photovoltaic cells, excitons are broken up into free electron-hole pairs by effective fields. The effective fields are set up by creating a heterojunction between two dissimilar materials. In organic photovoltaics, effective fields break up excitons by causing the electron to fall from the conduction band of the absorber to the conduction band of the acceptor molecule. It is necessary that the acceptor material has a conduction band edge that is lower than that of the absorber material.

Polymer solar cells usually consist of an electron- or hole-blocking layer on top of an indium tin oxide (ITO) conductive glass followed by electron donor and an electron acceptor (in the case of bulk heterojunction solar cells), a hole or electron blocking layer, and metal electrode on top. The nature and order of the blocking layers – as well as the nature of the metal electrode – depends on whether the cell follows a regular or an inverted device architecture. In an inverted cell, the electric charges exit the device in the opposite direction as in a normal device because the positive and negative electrodes are reversed. Inverted cells can utilize cathodes out of a more suitable material; inverted OPVs enjoy longer lifetimes than regularly structured OPVs, but they typically don't reach efficiencies as high as regular OPVs.

In bulk heterojunction polymer solar cells, light generates excitons. Subsequent charge separation in the interface between an electron donor and acceptor blend within the device's active layer. These charges then transport to the device's electrodes where the charges flow outside the cell, perform work and then re-enter the device on the opposite side. The cell's efficiency is limited by several factors, especially non-geminate recombination. Hole mobility leads to faster conduction across the active layer.

Organic photovoltaics are made of electron donor and electron acceptor materials rather than semiconductor p-n junctions. The molecules forming the electron donor region of organic PV cells, where exciton electron-hole pairs are generated, are generally conjugated polymers possessing delocalized π electrons that result from carbon p orbital hybridization. These π electrons can be excited by light in or near the visible part of the spectrum from the molecule's highest occupied molecular orbital (HOMO) to the lowest unoccupied molecular orbital (LUMO), denoted by a π -π^* transition. The energy bandgap between these orbitals determines which wavelength(s) of light can be absorbed.

Unlike in an inorganic crystalline PV cell material, with its band structure and delocalized electrons, excitons in organic photovoltaics are strongly bound with an energy between 0.1 and 1.4 eV. This strong binding occurs because electronic wave functions in organic molecules are more localized, and electrostatic attraction can thus keep the electron and hole together as an exciton. The electron and hole can be dissociated by providing an interface across which the chemical potential of electrons decreases. The material that absorbs the photon is the donor, and the material acquiring the electron is called the acceptor. In the polymer chain is the donor and the fullerene is the acceptor. Even after dissociation, the electron and hole may still be joined as a "geminate pair", and an electric field is then required to separate them. The electron and hole must be collected at contacts. If charge carrier mobility is insufficient, the carriers will not reach the contacts, and instead recombine at trap sites or remain in the device as undesirable space charges that oppose the flow of new carriers. The latter problem can occur if electron and hole mobilities are not matched. In that case, space-charge limited photocurrent (SCLP) hampers device performance.

Organic photovoltaics can be fabricated with an active polymer and a fullerene-based electron acceptor. Illumination of this system by visible light leads to electron transfer from the polymer to a fullerene molecule. As a result, the formation of a photoinduced quasiparticle, or polaron (P^+), occurs on the polymer chain and the fullerene becomes a radical anion (C^-_{60}). Polarons are highly mobile and can diffuse away.

Junction Types

The simplest organic PV device features a planar heterojunction. A film of organic active material (polymer or small molecule), of electron donor or electron acceptor type is sandwiched between contacts. Excitons created in the active material may diffuse before recombining and separate, hole and electron diffusing to its specific collecting electrode. Because charge carriers have diffusion lengths of just 3–10 nm in typical amorphous organic semiconductors, planar cells must be thin, but the thin cells absorb light less well. Bulk heterojunctions (BHJs) address this shortcoming. In a BHJ, a blend of electron donor and acceptor materials is cast as a mixture, which then phase-separates. Regions of each material in the device are separated by only several nanometers, a distance suited for carrier diffusion. BHJs require sensitive control over materials morphology on the nanoscale. Important variables include materials, solvents and the donor-acceptor weight ratio.

The next logical step beyond BHJs are ordered nanomaterials for solar cells, or ordered heterojunctions (OHJs). OHJs minimize the variability associated with BHJs. OHJs are generally hybrids of ordered inorganic materials and organic active regions. For example, a photovoltaic polymer can be deposited into pores in a ceramic such as TiO_2. Since holes still must diffuse the length of the pore through the polymer to a contact, OHJs suffer similar thickness limitations. Mitigating the hole mobility bottleneck is key to further enhancing device performance of OHJ's.

Single Layer

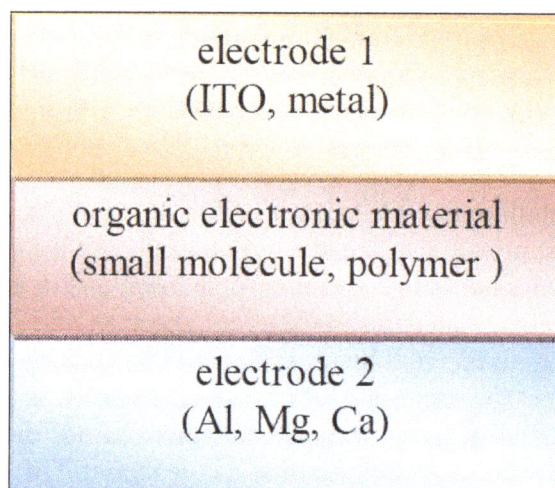

| electrode 1 |
| (ITO, metal) |
| organic electronic material |
| (small molecule, polymer) |
| electrode 2 |
| (Al, Mg, Ca) |

Sketch of a single layer organic photovoltaic cell.

Single layer organic photovoltaic cells are the simplest form. These cells are made by sandwiching a layer of organic electronic materials between two metallic conductors, typically a layer of indium tin oxide (ITO) with high work function and a layer of low work function metal such as Aluminum, Magnesium or Calcium. The basic structure of such a cell is illustrated in figure.

The difference of work function between the two conductors sets up an electric field in the organic layer. When the organic layer absorbs light, electrons will be excited to the LUMO and leave holes in the HOMO, thereby forming excitons. The potential created by the different work functions helps to split the exciton pairs, pulling electrons to the positive electrode (an electrical conductor used to make contact with a non-metallic part of a circuit) and holes to the negative electrode.

Examples:

In 1958 the photovoltaic effect or the creation of voltage of a cell based on magnesium phthalocyanine (MgPc)—a macrocyclic compound having an alternating nitrogen atom-carbon atom ring structure—was discovered to have a photovoltage of 200 mV. An Al/MgPc/Ag cell obtained photovoltaic efficiency of 0.01% under illumination at 690 nm.

Conjugated polymers were also used in this type of photovoltaic cell. One device used polyacetylene as the organic layer, with Al and graphite, producing an open circuit voltage of 0.3 V and a charge collection efficiency of 0.3%. An Al/poly(3-nethyl-thiophene)/Pt cell had an external quantum yield of 0.17%, an open circuit voltage of 0.4 V and a fill factor of 0.3. An ITO/PPV/Al cell showed an open circuit voltage of 1 V and a power conversion efficiency of 0.1% under white-light illumination.

Issues

Single layer organic solar cells do not work well. They have low quantum efficiencies (<1%) and low power conversion efficiencies (<0.1%). A major problem with them is that the electric field resulting from the difference between the two conductive electrodes is seldom sufficient to split the excitons. Often the electrons recombine with the holes without reaching the electrode.

Bilayer

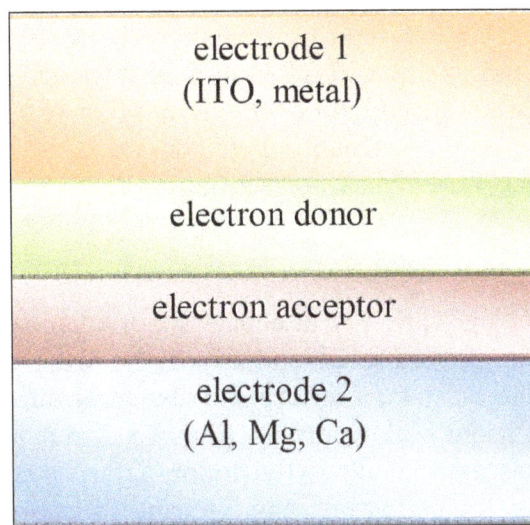

electrode 1
(ITO, metal)

electron donor

electron acceptor

electrode 2
(Al, Mg, Ca)

Sketch of a multilayer organic photovoltaic cell.

Bilayer cells contain two layers in between the conductive electrodes. The two layers have different electron affinity and ionization energies, therefore electrostatic forces are generated at the interface between the two layers. Light must create excitons in this small charged region for an efficient charge separation and collecting. The materials are chosen to make the differences large enough that these local electric fields are strong, which splits excitons much more efficiently than single layer photovoltaic cells. The layer with higher electron affinity and ionization potential is the electron acceptor, and the other layer is the electron donor. This structure is also called a planar donor-acceptor heterojunction.

Examples:

C_{60} has high electron affinity, making it a good acceptor. A C_{60}/MEH-PPV double layer cell had a relatively high fill factor of 0.48 and a power conversion efficiency of 0.04% under monochromatic illumination. PPV/C_{60} cells displayed a monochromatic external quantum efficiency of 9%, a power conversion efficiency of 1% and a fill factor of 0.48.

Perylene derivatives display high electron affinity and chemical stability. A layer of copper phthalocyanine (CuPc) as electron donor and perylene tetracarboxylic derivative as electron acceptor, fabricating a cell with a fill factor as high as 0.65 and a power conversion efficiency of 1% under simulated AM2 illumination. Halls et al. fabricated a cell with a layer of bis(phenethylimido) perylene over a layer of PPV as the electron donor. This cell had peak external quantum efficiency of 6% and power conversion efficiency of 1% under monochromatic illumination, and a fill factor of up to 0.6.

Issues

The diffusion length of excitons in organic electronic materials is typically on the order of 10 nm. In order for most excitons to diffuse to the interface of layers and split into carriers, the layer thickness should be in the same range as the diffusion length. However, a polymer layer typically needs a thickness of at least 100 nm to absorb enough light. At such a large thickness, only a small fraction of the excitons can reach the heterojunction interface.

Discrete Heterojunction

A three-layer (two acceptor and one donor) fullerene-free stack achieved a conversion efficiency of 8.4%. The implementation produced high open-circuit voltages and absorption in the visible spectra and high short-circuit currents. Quantum efficiency was above 75% between 400 nm and 720 nm wavelengths, with an open-circuit voltage around 1 V.

Bulk Heterojunction

Bulk heterojunctions have an absorption layer consisting of a nanoscale blend of donor and acceptor materials. The domain sizes of this blend are on the order of nanometers, allowing for excitons with short lifetimes to reach an interface and dissociate due to the large donor-acceptor interfacial area. However, efficient bulk heterojunctions need to maintain large enough domain sizes to form a percolating network that allows the donor materials to reach the hole transporting electrode and the acceptor materials to reach the electron transporting electrode. Without this percolating network, charges might be trapped in a donor or acceptor rich domain and undergo recombination. Bulk heterojunctions have an advantage over layered photoactive structures because they can be made thick enough for effective photon absorption without the difficult processing involved in orienting a layered structure while retaining similar level of performances.

Bulk heterojunctions are most commonly created by forming a solution containing the two components, casting (e.g. drop casting and spin coating) and then allowing the two phases to separate, usually with the assistance of an annealing step. The two components will self-assemble into an interpenetrating network connecting the two electrodes. They are normally composed of a conjugated molecule based donor and fullerene based acceptor. The nanostructural morphology of bulk heterojunctions tends to be difficult to control, but is critical to photovoltaic performance.

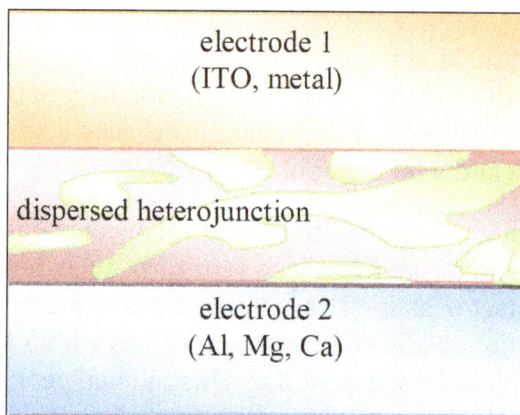

Sketch of a dispersed junction photovoltaic cell.

After the capture of a photon, electrons move to the acceptor domains, then are carried through the device and collected by one electrode, and holes move in the opposite direction and collected at the other side. If the dispersion of the two materials is too fine, it will result in poor charge transfer through the layer.

Most bulk heterojunction cells use two components, although three-component cells have been explored. The third component, a secondary p-type donor polymer, acts to absorb light in a different region of the solar spectrum. This in theory increases the amount of absorbed light. These ternary cells operate through one of three distinct mechanisms: charge transfer, energy transfer or parallel-linkage.

In charge transfer, both donors contribute directly to the generation of free charge carriers. Holes pass through only one donor domain before collection at the anode. In energy transfer, only one donor contributes to the production of holes. The second donor acts solely to absorb light, transferring extra energy to the first donor material. In parallel linkage, both donors produce excitons independently, which then migrate to their respective donor/acceptor interfaces and dissociate.

Examples:

Fullerenes such as C_{60} and its derivatives are used as electron acceptor materials in bulk heterojunction photovoltaic cells. A cell with the blend of MEH-PPV and a methano-functionalized C_{60} derivative as the heterojunction, ITO and Ca as the electrodes showed a quantum efficiency of 29% and a power conversion efficiency of 2.9% under monochromatic illumination. Replacing MEH-PPV with P3HT produced a quantum yield of 45% under a 10 V reverse bias. Further advances in modifying the electron acceptor has resulted in a device with a power conversion efficiency of 10.61% with a blend of $PC_{71}BM$ as the electron acceptor and PTB7-Th as the electron donor.

Polymer/polymer blends are also used in dispersed heterojunction photovoltaic cells. A blend of CN-PPV and MEH-PPV with Al and ITO as the electrodes, yielded peak monochromatic power conversion efficiency of 1% and fill factor of 0.38.

Dye sensitized photovoltaic cells can also be considered important examples of this type.

Issues

Fullerenes such as $PC_{71}BM$ are often the electron acceptor materials found in high performing bulk heterojunction solar cells. However, these electron acceptor materials very weakly absorb visible light, decreasing the volume fraction occupied by the strongly absorbing electron donor material. Furthermore, fullerenes have poor electronic tunability, resulting in restrictions placed on the development of conjugated systems with more appealing electronic structures for higher voltages. Recent research has been done on trying to replace these fullerenes with organic molecules that can be electronically tuned and contribute to light absorption.

Graded Heterojunction

The electron donor and acceptor are mixed in such a way that the gradient is gradual. This architecture combines the short electron travel distance in the dispersed heterojunction with the advantage of the charge gradient of the bilayer technology.

Examples: A cell with a blend of CuPc and C_{60} showed a quantum efficiency of 50% and a power conversion efficiency of 2.1% using 100 mW/cm² simulated AM1.5G solar illumination for a graded heterojunction.

Continuous Junction

Similar to the graded heterojunction the continuous junction concept aims at realizing a gradual transition from an electron donor to an electron acceptor. However, the acceptor material is prepared directly from the donor polymer in a post-polymerization modification step.

Production

Since its active layer largely determines device efficiency, this component's morphology received much attention.

If one material is more soluble in the solvent than the other, it will deposit first on top of the substrate, causing a concentration gradient through the film. This has been demonstrated for poly-3-hexyl thiophene (P3HT), phenyl-C_{61}-butyric acid methyl ester (PCBM) devices where the PCBM tends to accumulate towards the device's bottom upon spin coating from ODCB solutions. This effect is seen because the more soluble component tends to migrate towards the "solvent rich" phase during the coating procedure, accumulating the more soluble component towards the film's bottom, where the solvent remains longer. The thickness of the generated film affects the phases segregation because the dynamics of crystallization and precipitation are different for more concentrated solutions or faster evaporation rates (needed to build thicker devices). Crystalline P3HT enrichment closer to the hole-collecting electrode can only be achieved for relatively thin (100 nm) P3HT/PCBM layers.

The gradients in the initial morphology are then mainly generated by the solvent evaporation rate and the differences in solubility between the donor and acceptor inside the blend. This dependence on solubility has been clearly demonstrated using fullerene derivatives and P3HT. When using solvents which evaporate at a slower rate (as chlorobenzene (CB) or dichlorobenzene (DCB)) you can get larger degrees of vertical separation or aggregation while solvents that evaporate quicker produce a much less effective vertical separation. Larger solubility gradients should lead to more effective vertical separation while smaller gradients should lead to more homogeneous films. These two effects were verified on P3HT:PCBM solar cells.

The solvent evaporation speed as well as posterior solvent vapor or thermal annealing procedures were also studied. Blends such as P3HT:PCBM seem to benefit from thermal annealing procedures, while others, such as PTB7:PCBM, seem to show no benefit. In P3HT the benefit seems to come from an increase of crystallinity of the P3HT phase which is generated through an expulsion of PCBM molecules from within these domains. This has been demonstrated through studies of PCBM miscibility in P3HT as well as domain composition changes as a function of annealing times.

The above hypothesis based on miscibility does not fully explain the efficiency of the devices as solely pure amorphous phases of either donor or acceptor materials never exist within bulk heterojunction devices. A 2010 paper suggested that current models that assume pure phases and discrete interfaces might fail given the absence of pure amorphous regions. Since current models assume phase separation at interfaces without any consideration for phase purity, the models might need to be changed.

The thermal annealing procedure varies depending on precisely when it is applied. Since vertical species migration is partly determined by the surface tension between the active layer and either air or another layer, annealing before or after the deposition of additional layers (most often the metal cathode) affects the result. In the case of P3HT:PCBM solar cells vertical migration is improved when cells are annealed after the deposition of the metal cathode.

Donor or acceptor accumulation next to the adjacent layers might be beneficial as these accumulations can lead to hole or electron blocking effects which might benefit device performance. In 2009 the difference in vertical distribution on P3HT:PCBM solar cells was shown to cause problems with electron mobility which ends up with the yielding of very poor device efficiencies. Simple changes to device architecture – spin coating a thin layer of PCBM on top of the P3HT – greatly enhance cell reproducibility, by providing reproducible vertical separation between device components. Since higher contact between the PCBM and the cathode is required for better efficiencies, this largely increases device reproducibility.

According to neutron scattering analysis, P3HT:PCBM blends have been described as "rivers" (P3HT regions) interrupted by "streams" (PCBM regions).

Solvent Effects

Conditions for spin coating and evaporation affect device efficiency. Solvent and additives influence donor-acceptor morphology. Additives slow down evaporation, leading to more crystalline polymers and thus improved hole conductivities and efficiencies. Typical additives include 1,8-octanedithiol, ortho-dichlorobenzene, 1,8-diiodooctane (DIO), and nitrobenzene. The DIO effect was attributed to the selective solubilization of PCBM components, modifies fundamentally the average hopping distance of electrons, and thus improves electron mobility. Additives can also lead to big increases in efficiency for polymers. For HXS-1/PCBM solar cells, the effect was correlated with charge generation, transport and shelf-stability. Other polymers such as PTTBO also benefit significantly from DIO, achieving PCE values of more than 5% from around 3.7% without the additive.

Polymer Solar Cells fabricated from chloronaphthalene (CN) as a co-solvent enjoy a higher efficiency than those fabricated from the more conventional pure chlorobenzene solution. This is because the donor-acceptor morphology changes, which reduces the phase separation between donor polymer and fullerene. As a result, this translates into high hole mobilities. Without co-solvents, large domains of fullerene form, decreasing photovoltaic performance of the cell due to polymer aggregation in solution. This morphology originates from the liquid-liquid phase separation during drying; solve evaporation causes the mixture to enter into the spinodal region, in which there are significant thermal fluctuations. Large domains prevent electrons from being collected efficiently (decreasing PCE).

Small differences in polymer structure can also lead to significant changes in crystal packing that inevitably affect device morphology. PCPDTBT differs from PSBTBT caused by the difference in bridging atom between the two polymers (C vs. Si), which implies that better morphologies are achievable with PCPDTBT:PCBM solar cells containing additives as opposed to the Si system which achieves good morphologies without help from additional substances.

Self-assembled Cells

Supramolecular chemistry was investigated, using donor and acceptor molecules that assemble upon spin casting and heating. Most supramolecular assemblies employ small molecules. Donor and acceptor domains in a tubular structure appear ideal for organic solar cells.

Diblock polymers containing fullerene yield stable organic solar cells upon thermal annealing. Solar cells with pre-designed morphologies resulted when appropriate supramolecular interactions are introduced.

Progress on BCPs containing polythiophene derivatives yield solar cells that assemble into well defined networks. This system exhibits a PCE of 2.04%. Hydrogen bonding guides the morphology.

Device efficiency based on co-polymer approaches have yet to cross the 2% barrier, whereas bulk-heterojunction devices exhibit efficiencies >7% in single junction configurations.

Fullerene-grafted rod-coil block copolymers have been used to study domain organization.

Supramolecular approaches to organic solar cells provide understanding about the macromolecular forces that drive domain separation.

Infrared Polymer Cells

Infrared cells preferentially absorb light in the infrared range rather than visible wavelengths. As of 2012, such cells can be made nearly 70% transparent to visible light. The cells allegedly can be made in high volume at low cost using solution processing. Infrared polymer cells can be used as add-on components of portable electronics, smart windows, and building-integrated photovoltaics.The cells employ silver nanowire/titanium dioxide composite films as the top electrode, replacing conventional opaque metal electrodes. With this combination, 4% power-conversion efficiency was achieved.

Near-infrared Polymer solar cells based on a copolymer of naphthodithiophene diimide and bithiophene (PNDTI-BT-DT) are also being fabricated in combination with PTB7 as an electron donor. Both PNDTI-BT-DT and PTB7 formed a crystalline structure in the blend films similar to in the pristine films, leading to the efficient charge generation contributed from both polymers.

Typical Current-voltage Behavior and Power Conversion Efficiency

Organic photovoltaics, similar to inorganic photovoltaics, are generally characterized through current-voltage analysis. This analysis provides multiple device metrics values that are used to understand device performance. One of the most crucial metrics is the Power Conversion Efficiency (PCE).

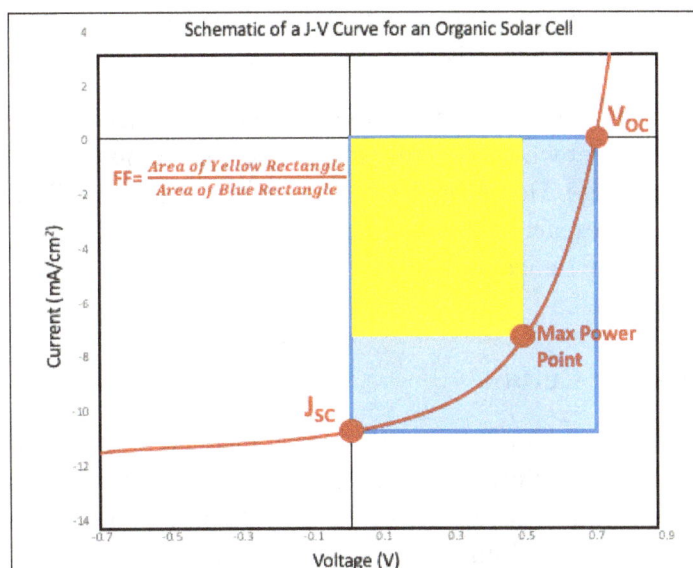

PCE (η) is proportional to the product of the short-circuit current (J_{SC}), the open circuit voltage (V_{OC}), and the fill factor (FF), all of which can be determined from a current-voltage curve.

$$\eta = \frac{V_{OC} \times J_{SC} \times FF}{P_{in}}$$

Where P_{in} is the incident solar power.

The short circuit current (Jsc), is the maximum photocurrent generation value. It corresponds to the y-intercept value of standard current-voltage curve in which current is plotted along the y-axis and voltage is plotted along the x-axis. Within organic solar cells, the short circuit current can be impacted by a variety of material factors. These include the mobility of charge carriers, the optical absorption profile and general energetic driving forces that lead to a more efficient extraction of charge carriers.

The open circuit voltage (Voc) is the voltage when there is no current running through the device. This corresponds to the x-intercept on a current-voltage curve. Within bulk heterojunction organic photovoltaic devices, this value is highly dependent on HOMO and LUMO energy levels and work functions for the active layer materials.

Since power is the product of voltage and current, the maximum power point occurs when the product between voltage and current is maximized.

The fill factor, FF, can be thought of as the "squareness" of a current voltage curve. It is the quotient of the maximum power value and the product of the open circuit voltage and short circuit current. This is shown in the image above as the ratio of the area of the yellow rectangle to the greater blue rectangle. For organic photovoltaics, this fill factor is essentially a measure of how efficiently generated charges are extracted from the device. This can be thought of as a "competition" between charges transporting through the device, and charges that recombine.

A major issue surrounding polymer solar cells is the low Power Conversion Efficiency (PCE) of fabricated cells. In order to be considered commercially viable, PSCs must be able to achieve at least

10–15% efficiency—this is already much lower than inorganic PVs. However, due to the low cost of polymer solar cells, a 10–15% efficiency is commercially viable.

Recent advances in polymer solar cell performance have resulted from compressing the bandgap to enhance short-circuit current while lowering the Highest Occupied Molecular Orbital (HOMO) to increase open-circuit voltage. However, PSCs still suffer from low fill factors (typically below 70%). However, as of 2013, researchers have been able to fabricate PSCs with fill factors of over 75%. Scientists have been able to accomplish via an inverted BHJ and by using nonconventional donor/acceptor combinations.

Modeling Organic Solar Cells

As discussed above, organic semiconductors are highly disordered materials with no long range order. This means that the conduction band and valance band edges are not well defined. Furthermore, this physical and energetic disorder generates trap states in which photogenerated electrons and holes can become trapped and then eventually recombine.

Key to accurately describing organic solar cells in a device model is to include carrier trapping and recombination via trap states. A commonly used approach is to use an effective medium model, where by standard drift diffusion equations are used to describe transport across the device. Then, an exponential tail of trap states is introduced which decays into the band gap from the mobility edges. To describe capture/escape from these trap states the Shockley–Read–Hall (SRH) can be used. The Shockley-Read-Hall mechanism has been shown able to reproduce polymer:fullerene device behavior in both time domain and steady state.

Challenges and Progress

Difficulties associated with organic photovoltaic cells include their low external quantum efficiency (up to 70%) compared to inorganic photovoltaic devices, despite having good internal quantum efficiency; this is due to insufficient absorption with active layers on the order of 100 nanometers. Instabilities against oxidation and reduction, recrystallization and temperature variations can also lead to device degradation and decreased performance over time. This occurs to different extents for devices with different compositions, and is an area into which active research is taking place.

Other important factors include the exciton diffusion length, charge separation and charge collection which are affected by the presence of impurities.

Charge Carrier Mobility and Transport

Especially for bulk heterojunction solar cells, understanding charge carrier transport is vital in improving the efficiencies of organic photovoltaics. Currently, bulk heterojunction devices have imbalanced charge-carrier mobility, with the hole mobility being at least an order of magnitude lower than that of the electron mobility; this results in space charge build-up and a decrease in the fill factor and power conversation efficiency of a device. Due to having low mobility, efficient bulk heterojunction photovoltaics have to be designed with thin active layers to avoid recombination of the charge carriers, which is detrimental to absorption and scalability in processing. Simulations have demonstrated that in order to have an bulk heterojunction solar cell with a fill factor above

0.8 and external quantum efficiency above 90%, there needs to be balanced charge carrier mobility to reduce a space charge effect, as well as an increase in charge carrier mobility and/or a decrease in the bimolecular recombination rate constant.

Effect of Film Morphology

As described above, dispersed heterojunctions of donor-acceptor organic materials have high quantum efficiencies compared to the planar hetero-junction, because in dispersed heterojunctions it is more likely for an exciton to find an interface within its diffusion length. Film morphology can also have a drastic effect on the quantum efficiency of the device. Rough surfaces and the presence of voids can increase the series resistance and also the chance of short-circuiting. Film morphology and, as a result, quantum efficiency can be improved by annealing of a device after covering it by a ~1000 Å thick metal cathode. Metal film on top of the organic film applies stresses on the organic film, which helps to prevent the morphological relaxation in the organic film. This gives more densely packed films and at the same time allows the formation of phase-separated interpenetrating donor-acceptor interface inside the bulk of organic thin film.

Controlled Growth Heterojunction

Charge separation occurs at the donor-acceptor interface. Whilst traveling to the electrode, a charge can become trapped and/or recombine in a disordered interpenetrating organic material, resulting in decreased device efficiency. Controlled growth of the heterojunction provides better control over positions of the donor-acceptor materials, resulting in much greater power efficiency (ratio of output power to input power) than that of planar and highly disoriented hetero-junctions. Thus, the choice of suitable processing parameters in order to better control the structure and film morphology is highly desirable.

Progress in Growth Techniques

Mostly organic films for photovoltaic applications are deposited by spin coating and vapor-phase deposition. However each method has certain draw backs, spin coating technique can coat larger surface areas with high speed but the use of solvent for one layer can degrade the already existing polymer layer. Another problem is related with the patterning of the substrate for device as spin-coating results in coating the entire substrate with a single material.

Vacuum Thermal Evaporation

Another deposition technique is vacuum thermal evaporation (VTE) which involves the heating of an organic material in vacuum. The substrate is placed several centimeters away from the source so that evaporated material may be directly deposited onto the substrate, as shown in figure (a). This method is useful for depositing many layers of different materials without chemical interaction between different layers. However, there are sometimes problems with film-thickness uniformity and uniform doping over large-area substrates. In addition, the materials that deposit on the wall of the chamber can contaminate later depositions. This "line of sight" technique also can create holes in the film due to shadowing, which causes an increase in the device series-resistance and short circuit.

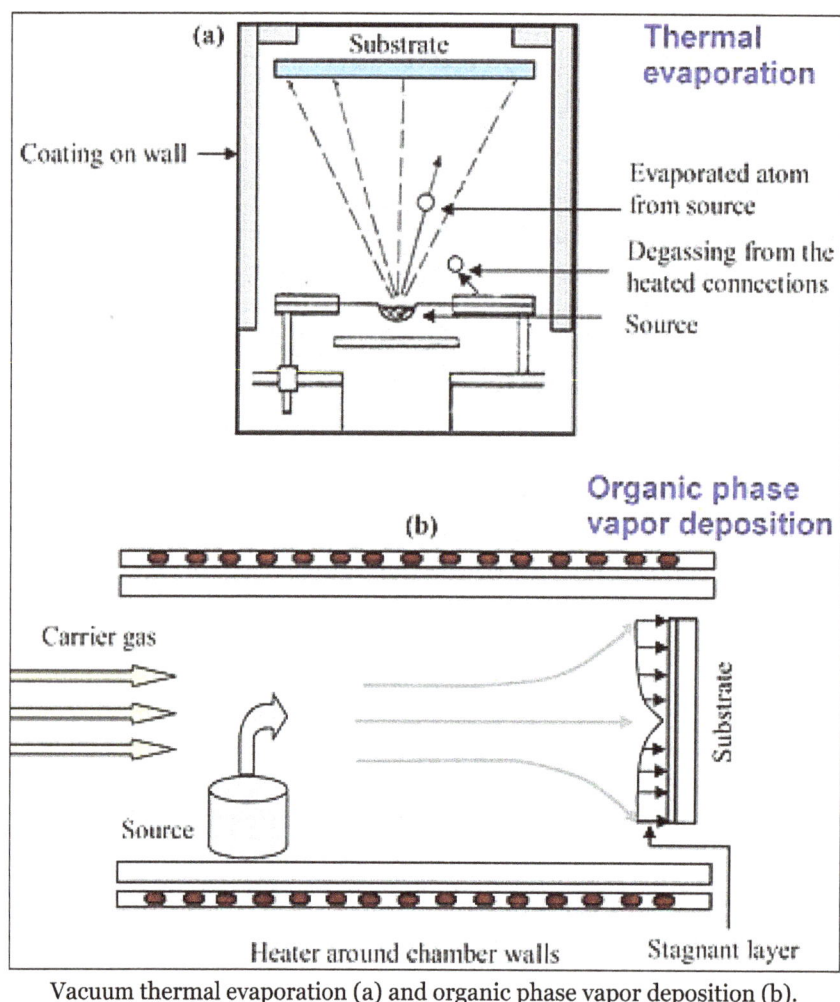

Vacuum thermal evaporation (a) and organic phase vapor deposition (b).

Organic Vapor Phase Deposition

Organic vapor phase deposition (OVPD, figure (b)) allows better control of the structure and morphology of the film than vacuum thermal evaporation. The process involves evaporation of the organic material over a substrate in the presence of an inert carrier gas. The resulting film morphology can be tuned by changing the gas flow rate and the source temperature. Uniform films can be grown by reducing the carrier gas pressure, which will increase the velocity and mean free path of the gas, and as a result boundary layer thickness decreases. Cells produced by OVPD do not have issues related with contaminations from the flakes coming out of the walls of the chamber, as the walls are warm and do not allow molecules to stick to and produce a film upon them.

Another advantage over VTE is the uniformity in evaporation rate. This occurs because the carrier gas becomes saturated with the vapors of the organic material coming out of the source and then moves towards the cooled substrate, figure (b). Depending on the growth parameters (temperature of the source, base pressure and flux of the carrier gas) the deposited film can be crystalline or amorphous in nature. Devices fabricated using OVPD show a higher short-circuit current density than that of devices made using VTE. An extra layer of donor-acceptor hetero-junction at the top of the cell may block excitons, whilst allowing conduction of electron; resulting in improved cell efficiency.

Organic Solar Ink

Organic solar ink is able to deliver higher performance in *fluorescent* lighting conditions in comparison to amorphous silicon solar cells, and said to have a 30% to 40% increase in indoor power density in comparison to the standard organic solar technology.

Light Trapping

Various type of components are applied to increase light trapping (Light in-coupling) effects in thin organic solar cells. In addition to the flexibility of organic solar cells, by using flexible electrodes and substrates instead of ITO and glass respectively, fully flexible organic solar cells can be produced. By these use of flexible substrates and substrates, easier methods to provide light trapping effects to OPVs are introduced such as polymer electrodes with embedded scattering particles, nano imprinted polymer electrodes, patterned PET substrates and even optical display film commercialized for liquid crystal displays (LCD) as substrates. Much research will be taken for enhancing the performance of OPVs with the merit of easy light trapping structures processing.

Use in Tandem Photovoltaics

Recent research and study has been done in utilizing an organic solar cell as the top cell in a hybrid tandem solar cell stack. Because organic solar cells have a higher band gap than traditional inorganic photovoltaics like silicon or CIGS, they can absorb higher energy photons without losing much of the energy due to thermalization, and thus operate at a higher voltage. The lower energy photons and higher energy photons that are unabsorbed pass through the top organic solar cell and are then absorbed by the bottom inorganic cell. Organic solar cells are also solution processible at low temperatures with a low cost of 10 dollars per square meter, resulting in a printable top cell that improves the overall efficiencies of existing, inorganic solar cell technologies. Much research has been done to enable the formation of such a hybrid tandem solar cell stack, including research in the deposition of semi-transparent electrodes that maintain low contact resistance while having high transparency.

Recent Directions for Bulk Heterojunction Materials Research

One major area of current research is the use of non-fullerene acceptors. While fullerene acceptors have been the standard for most organic photovoltaics due to their compatibility within bulk heterojunction cell designs as well as their good transport properties, they do have some fallbacks that are leading researchers to attempt to find alternatives. Some negatives of fullerene acceptors include their instability, that they are somewhat limited in energy-tunability and they have poor optical absorption. Researchers have developed small molecule acceptors that due to their good energy tunability, can exhibit high open circuit voltages. However, there are still major challenges with non-fullerene acceptors, including the low charge carrier mobilities of small molecule acceptors, and that the sheer number of possible molecules is overwhelming for the research community.

Small molecules are also being heavily researched to act as donor materials, potentially replacing polymeric donors. Since small molecules do not vary in molecular weights the way polymers do, they would require less purification steps and are less susceptible to macromolecule defects and kinks that can create trap states leading to recombination. Recent research has shown that high-performing small molecular donor structures tend to have planar 2-D structures and can aggregate or self assemble. Sine performance of these devices is highly depended on active layer morphology, present research is continuing to investigate small molecule possibilities, and optimize device morphology through processes such as annealing for various materials.

Hybrid Solar Cell

Hybrid solar cells combine advantages of both organic and inorganic semiconductors. Hybrid photovoltaics have organic materials that consist of conjugated polymers that absorb light as the donor and transport holes. Inorganic materials in hybrid cells are used as the acceptor and electron transporter in the structure. The hybrid photovoltaic devices have a potential for not only low-cost by roll-to-roll processing but also for scalable solar power conversion.

Solar cells are devices that convert sunlight into electricity by the photovoltaic effect. Electrons in a solar cell absorb photon energy in sunlight which excites them to the conduction band from the valence band. This generates a hole-electron pair, which is separated by a potential barrier (such as a p-n junction), and induces a current. Organic solar cells use organic materials in their active layers. Molecular, polymer, and hybrid organic photovoltaics are the main kinds of organic photovoltaic devices currently studied.

In hybrid solar cells, an organic material is mixed with a high electron transport material to form the photoactive layer. The two materials are assembled together in a heterojunction-type photoactive layer, which can have a greater power conversion efficiency than a single material. One of the materials acts as the photon absorber and exciton donor. The other material facilitates exciton dissociation at the junction. Charge is transferred and then separated after an exciton created in the donor is delocalized on a donor-acceptor complex.

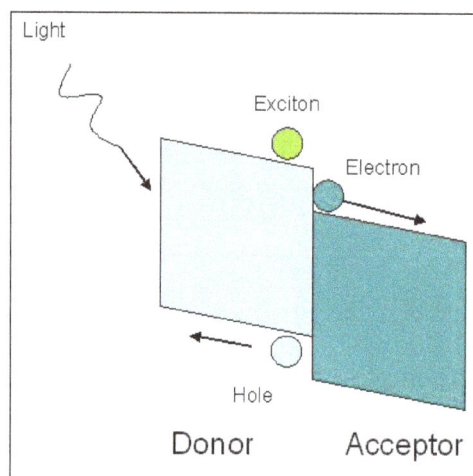

Energy diagram of the donor and acceptor. The conduction band of the acceptor is lower than the LUMO of the polymer, allowing for transfer of the electron.

The acceptor material needs a suitable energy offset to the binding energy of the exciton to the absorber. Charge transfer is favorable if the following condition is satisfied:

$$E_A^A - E_A^D > U_D$$

where superscripts A and D refer to the acceptor and donor respectively, E_A is the electron affinity, and U the coulombic binding energy of the exciton on the donor. An energy diagram of the interface is shown in figure. In commonly used photovoltaic polymers such as MEH-PPV, the exciton binding energy ranges from 0.3 eV to 1.4 eV.

The energy required to separate the exciton is provided by the energy offset between the LUMOs or conduction bands of the donor and acceptor. After dissociation, the carriers are transported to the respective electrodes through a percolation network.

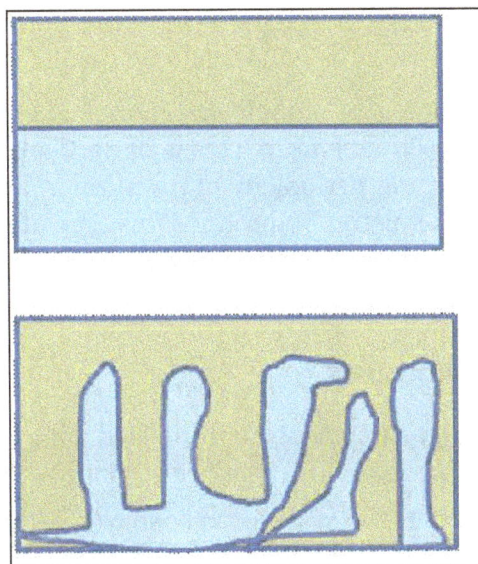

Two different structures of heterojunctions, a) phase separated bi-layer and b) bulk heterojunction. The bulk heterojunction allows for more interfacial contact between the two phases, which is beneficial for the nanoparticle-polymer compound as it provides more surface area for charge transfer.

The average distance an exciton can diffuse through a material before annihilation by recombination is the exciton diffusion length. This is short in polymers, on the order of 5–10 nanometers. The time scale for radiative and non-radiative decay is from 1 picosecond to 1 nanosecond. Excitons generated within this length close to an acceptor would contribute to the photocurrent.

To deal with the problem of the short exciton diffusion length, a bulk heterojunction structure is used rather than a phase-separated bilayer. Dispersing the particles throughout the polymer matrix creates a larger interfacial area for charge transfer to occur. Figure displays the difference between a bilayer and a bulk heterojunction.

Types of Interfaces and Structures

Controlling the interface of inorganic-organic hybrid solar cells can increase the efficiency of the cells. This increased efficiency can be achieved by increasing the interfacial surface area between the organic and the inorganic to facilitate charge separation and by controlling the nanoscale lengths and periodicity of each structure so that charges are allowed to separate and move toward the appropriate electrode without recombining. The three main nanoscale structures used are mesoporous inorganic films infused with electron-donating organic, alternating inorganic-organic lamellar structures, and nanowire structures.

Mesoporous Films

Mesoporous films have been used for a relatively high-efficiency hybrid solar cell. The structure of mesoporous thin film solar cells usually includes a porous inorganic that is saturated with organic surfactant. The organic absorbs light, and transfers electrons to the inorganic semiconductor (usually a transparent conducting oxide), which then transfers the electron to the electrode. Problems with these cells include their random ordering and the difficulty of controlling their nanoscale structure to promote charge conduction.

Ordered Lamellar Films

Recently, the use of alternating layers of organic and inorganic compounds has been controlled through electrodeposition-based self-assembly. This is of particular interest because it has been shown that the lamellar structure and periodicity of the alternating organic-inorganic layers can be controlled through solution chemistry. To produce this type of cell with practical efficiencies, larger organic surfactants that absorb more of the visible spectrum must be deposited between the layers of electron-accepting inorganic.

Films of Ordered Nanostructures

Researchers have been able to grow nanostructure-based solar cells that use ordered nanostructures like nanowires or nanotubes of inorganic surrounding by electron-donating organics utilizing self-organization processes. Ordered nanostructures offer the advantage of directed charge transport and controlled phase separation between donor and acceptor materials. The nanowire-based morphology offers reduced internal reflection, facile strain relaxation and increased defect tolerance. The ability to make single-crystalline nanowires on low-cost substrates such as aluminum foil and to relax strain in subsequent layers removes two more major cost hurdles associated with high-efficiency cells. There have been rapid increases in efficiencies of nanowire-based solar cells and they seem to be one of the most promising nanoscale solar hybrid technologies.

Fundamental Challenge Factors

Hybrid cell efficiency must be increased to start large-scale manufacturing. Three factors affect efficiency. First, the bandgap should be reduced to absorb red photons, which contain a significant fraction of the energy in the solar spectrum. Current organic photovoltaics have shown 70% of quantum efficiency for blue photons. Second, contact resistance between each layer in the device should be minimized to offer higher fill factor and power conversion efficiency. Third, charge-carrier mobility should be increased to allow the photovoltaics to have thicker active layers while minimizing carrier recombination and keeping the series resistance of the device low.

Types of Hybrid Solar Cells

Polymer–nanoparticle Composite

Nanoparticles are a class of semiconductor materials whose size in at least one dimension ranges from 1 to 100 nanometers, on the order of exciton wavelengths. This size control creates quantum confinement and allows for the tuning of optoelectronic properties, such as band gap and electron affinity. Nanoparticles also have a large surface area to volume ratio, which presents more area for charge transfer to occur.

The photoactive layer can be created by mixing nanoparticles into a polymer matrix. Solar devices based on polymer-nanoparticle composites most resemble polymer solar cells. In this case, the nanoparticles take the place of the fullerene based acceptors used in fully organic polymer solar cells. Hybrid solar cells based upon nanoparticles are an area of research interest because nanoparticles have several properties that could make them preferable to fullerenes, such as:

- Fullerenes are synthesized by a combination of a high temperature arc method and continuous gas-phase synthesis, which makes their production difficult and energy intensive. The colloidal synthesis of nanoparticles by contrast is a low temperature process.

- PCBM (a common fullerene acceptor) diffuses during long timespans or when exposed to heat, which can alter the morphology and lower the efficiency of a polymer solar cell. Limited testing of nanoparticle solar cells indicates they may be more stable over time.

- Nanoparticles are more absorbent than fullerenes, meaning more light can be theoretically absorbed in a thinner device.

- Nanoparticle size can affect absorption. This combined with the fact that there are many possible semiconducting nanoparticles allows for highly customizable bandgaps that can be easily tuned to certain frequencies, which would be advantageous in tandem solar cells.

- Nanoparticles with size near their Bohr radius can generate two excitons when struck by a sufficiently energetic photon.

Structure and Processing

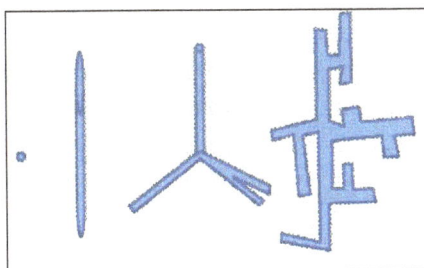

Four different structures of nanoparticles, which have at least 1 dimension in the 1–100 nm range, retaining quantum confinement. Left is a nanocrystal, next to it is nanorod, third is tetrapod, and right is hyperbranched.

For polymers used in this device, hole mobilities are greater than electron mobilities, so the polymer phase is used to transport holes. The nanoparticles transport electrons to the electrode.

The interfacial area between the polymer phase and the nanoparticles needs to be large. This is achieved by dispersing the particles throughout the polymer matrix. However, the nanoparticles need to be interconnected to form percolation networks for electron transport, which occurs by hopping events.

Efficiency is affected by aspect ratio, geometry, and volume fraction of the nanoparticles. Nanoparticle structures include nanocrystals, nanorods, and hyperbranched structures. contains a picture of each structure. Different structures change the conversion efficiency by effecting nanoparticle dispersion in the polymer and providing pathways for electron transport.

The nanoparticle phase is required to provide a pathway for the electrons to reach the electrode. By using nanorods instead of nanocrystals, the hopping event from one crystal to another can be avoided.

Fabrication methods include mixing the two materials in a solution and spin-coating it onto a substrate, and solvent evaporation (sol-gel). Most of these methods do not involve high-temperature processing. Annealing increases order in the polymer phase, increasing conductivity. However, annealing for too long causes the polymer domain size to increase, eventually making it larger than the exciton diffusion length, and possibly allowing some of the metal from the contact to diffuse into the photoactive layer, reducing the efficiency of the device.

Materials

Inorganic semiconductor nanoparticles used in hybrid cells include CdSe (size ranges from 6–20 nm), ZnO, TiO, and PbS. Common polymers used as photo materials have extensive conjugation and are also hydrophobic. Their efficiency as a photo-material is affected by the HOMO level position and the ionization potential, which directly affects the open circuit voltage and the stability in air. The most common polymers used are P3HT (poly (3-hexylthiophene)), and M3H-PPV (poly[2-methoxy, 5-(2'-ethyl-hexyloxy)-p-phenylenevinylene)]). P3HT has a bandgap of 2.1 eV and M3H-PPV has a bandgap of ~2.4 eV. These values correspond with the bandgap of CdSe, 2.10 eV. The electron affinity of CdSe ranges from 4.4 to 4.7 eV. When the polymer used is MEH-PPV, which has an electron affinity of 3.0 eV, the difference between the electron affinities is large enough to drive electron transfer from the CdSe to the polymer. CdSe also has a high electron mobility (600 $cm^2 \cdot V^{-1} \cdot s^{-1}$).

Performance Values

The highest demonstrated efficiency is 3.2%, based upon a PCPDTBT polymer donor and CdSe nanoparticle acceptor. The device exhibited a short circuit current of 10.1 $mA \cdot cm^{-2}$, an open circuit voltage of .68 V, and a fill factor of .51.

Challenges

Hybrid solar cells need increased efficiencies and stability over time before commercialization is feasible. In comparison to the 2.4% of the CdSe-PPV system, silicon photodevices have power conversion efficiencies greater than 20%.

Problems include controlling the amount of nanoparticle aggregation as the photolayer forms. The particles need to be dispersed in order to maximize interface area, but need to aggregate to form networks for electron transport. The network formation is sensitive to the fabrication conditions. Dead end pathways can impede flow. A possible solution is implementing ordered heterojunctions, where the structure is well controlled.

The structures can undergo morphological changes over time, namely phase separation. Eventually, the polymer domain size will be greater than the carrier diffusion length, which lowers performance.

Even though the nanoparticle bandgap can be tuned, it needs to be matched with the corresponding polymer. The 2.0 eV bandgap of CdSe is larger than an ideal bandgap of 1.4 for absorbance of light.

The nanoparticles involved are typically colloids, which are stabilized in solution by ligands. The ligands decrease device efficiency because they serve as insulators which impede interaction between the donor and nanoparticle acceptor as well as decreasing the electron mobility. Some, but not complete success has been had by exchanging the initial ligands for pyridine or another short chain ligand.

Hybrid solar cells exhibit material properties inferior to those of bulk silicon semiconductors. The carrier mobilities are much smaller than that of silicon. Electron mobility in silicon is 1000 $cm^2 \cdot V^{-1} \cdot s^{-1}$, compared to 600 $cm^2 \cdot V^{-1} \cdot s^{-1}$ in CdSe, and less than 10 $cm^2 \cdot V^{-1} \cdot s^{-1}$ in other quantum dot materials. Hole mobility in MEH-PPV is 0.1 $cm^2 \cdot V^{-1} \cdot s^{-1}$, while in silicon it is 450 $cm^2 \cdot V^{-1} \cdot s^{-1}$.

Carbon Nanotubes

Carbon nanotubes (CNTs) have high electron conductivity, high thermal conductivity, robustness, and flexibility. Field emission displays (FED), strain sensors, and field effect transistors (FET) using CNTs have been demonstrated. Each application shows the potential of CNTs for nanoscale devices and for flexible electronics applications. Photovoltaic applications have also been explored for this material.

Mainly, CNTs have been used as either the photo-induced exciton carrier transport medium impurity within a polymer-based photovoltaic layer or as the photoactive (photon-electron conversion) layer. Metallic CNT is preferred for the former application, while semiconducting CNT is preferred for the later.

Efficient Carrier Transport Medium

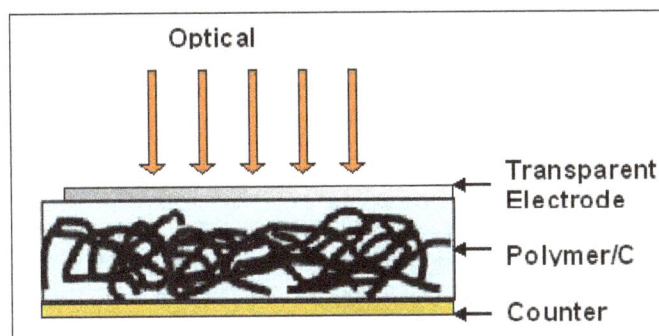

Device diagram for CNT as efficient carrier transport medium.

To increase the photovoltaic efficiency, electron-accepting impurities must be added to the photoactive region. By incorporating CNTs into the polymer, dissociation of the exciton pair can be accomplished by the CNT matrix. The high surface area (~1600 m^2/g) of CNTs offers a good opportunity for exciton dissociation. The separated carriers within the polymer-CNT matrix are transported by the percolation pathways of adjacent CNTs, providing the means for high carrier mobility and efficient charge transfer. The factors of performance of CNT-polymer hybrid photovoltaics are low compared to those of inorganic photovoltaics. SWNT in P3OT semiconductor polymer demonstrated open circuit voltage (V_{oc}) of below 0.94 V, with short circuit current (I_{sc}) of 0.12 mA/cm².

Metal nanoparticles may be applied to the exterior of CNTs to increase the exciton separation efficiency. The metal provides a higher electric field at the CNT-polymer interface, accelerating the exciton carriers to transfer them more effectively to the CNT matrix. In this case, V_{oc} = 0.3396 V and I_{sc} = 5.88 mA/cm². The fill factor is 0.3876%, and the white light conversion factor 0.775%.

Photoactive Matrix Layer

CNT may be used as a photovoltaic device not only as an add-in material to increase carrier transport, but also as the photoactive layer itself. The semiconducting single walled CNT (SWCNT) is a potentially attractive material for photovoltaic applications for the unique structural and electrical properties. SWCNT has high electric conductivity (100 times that of copper) and shows ballistic carrier transport, greatly decreasing carrier recombination. The bandgap of the SWCNT is inversely proportional to the tube diameter, which means that SWCNT may show multiple direct bandgaps matching the solar spectrum.

A strong built-in electric field in SWCNT for efficient photogenerated electron-hole pair separation has been demonstrated by using two asymmetrical metal electrodes with high and low work functions. The open circuit voltage (V_{oc}) is 0.28 V, with short circuit current (I_{sc}) 1.12 nA·cm⁻² with an incident light source of 8.8 W·cm⁻². The resulting white light conversion factor is 0.8%.

Challenges

Several challenges must be addressed for CNT to be used in photovoltaic applications. CNT degrades over time in an oxygen-rich environment. The passivation layer required to prevent CNT oxidation may reduce the optical transparency of the electrode region and lower the photovoltaic efficiency.

Challenges as Efficient Carrier Transport Medium

Additional challenges involve the dispersion of CNT within the polymer photoactive layer. The CNT is required to be well dispersed within the polymer matrix to form charge-transfer-efficient pathways between the excitons and the electrode.

Challenges as Photoactive Matrix Layer

Challenges of CNT for the photoactive layer include its lack of capability to form a p-n junction, due to the difficulty of doping certain segments of a CNT. (A p-n junction creates an internal built-in potential, providing a pathway for efficient carrier separation within the photovoltaic.) To overcome this difficulty, energy band bending has been done by the use of two electrodes of different work functions. A strong built-in electric field covering the whole SWCNT channel is formed for high-efficiency carrier separation. The oxidation issue with CNT is more critical for this application. Oxidized CNTs have a tendency to become more metallic, and so less useful as a photovoltaic material.

Dye-sensitized

Dye-sensitized solar cells consists of a photo-sensitized anode, an electrolyte, and a photo-electro-chemical system. Hybrid solar cells based on dye-sensitized solar cells are formed with inorganic materials (TiO_2) and organic materials.

Materials

Hybrid solar cells based on dye-sensitized solar cells are fabricated by dye-absorbed inorganic materials and organic materials. TiO_2 is the preferred inorganic material since this material is easy to synthesize and acts as a n-type semiconductor due to the donor-like oxygen vacancies. However, titania only absorbs a small fraction of the UV spectrum. Molecular sensitizers (dye molecules) attached to the semiconductor surface are used to collect a greater portion of the spectrum. In the case of titania dye-sensitized solar cells, a photon absorbed by a dye-sensitizer molecule layer induces electron injection into the conduction band of titania, resulting in current flow. However, short diffusion length (diffusivity, $D_n \leq 10^{-4} cm^2/s$) in titania dye-sensitized solar cells decrease the solar-to-energy conversion efficiency. To enhance diffusion length (or carrier lifetime), a variety of organic materials are attached to the titania.

Fabrication Scheme

Dye-sensitized Photoelectrochemical Cell (Grätzel Cell)

TiO_2 nanoparticles are synthesized in several tens of nanometer scales (~100 nm). In order to make a photovoltaic cell, molecular sensitizers (dye molecules) are attached to the titania surface. The dye-absorbed titania is finally enclosed by a liquid electrolyte. This type of dye-sensitized solar cell is also known as a Grätzel cell. Dye-sensitized solar cell has a disadvantage of a short diffusion length. Recently, supermolecular or multifunctional sensitizers have been investigated so as to enhance carrier diffusion length. For example, a dye chromophore has been modified by the addition of secondary electron donors. Minority carriers (holes in this case) diffuse to the attached electron donors to recombine. Therefore, electron-hole recombination is retarded by the physical separation between the dye–cation moiety and the TiO_2 surface, as shown in figure. Finally, this process raises the carrier diffusion length, resulting in the increase of carrier lifetime.

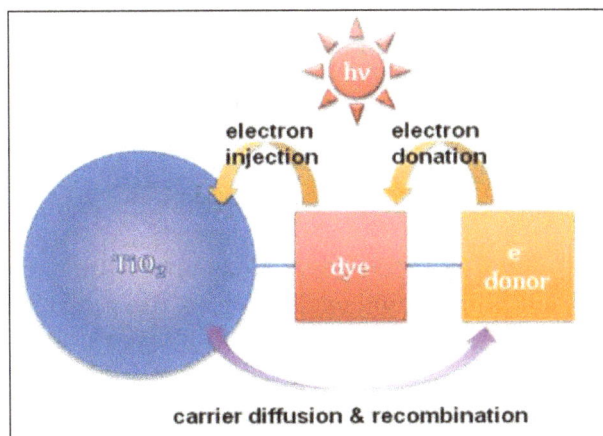

Schematic representation of electron-hole generation and recombination.

Solid-state Dye Sensitized Solar Cell

Mesoporous materials contain pores with diameters between 2 and 50 nm. A dye-sensitized meso-porous film of TiO_2 can be used for making photovoltaic cells and this solar cell is called a 'solid-state dye sensitized solar cell'. The pores in mesoporous TiO_2 thin film are filled with a solid hole-conducting material such as p-type semiconductors or organic hole conducting material. Replacing the liquid electrolyte in Grätzel's cells with a solid charge-transport material can be beneficial. The process of electron-hole generation and recombination is the same as Grätzel cells. Electrons are injected from photoexcited dye into the conduction band of titania and holes are transported by a solid charge transport electrolyte to an electrode. Many organic materials have been tested to obtain a high solar-to-energy conversion efficiency in dye synthesized solar cells based on mesoporous titania thin film.

Efficiency Factors

Efficiency factors demonstrated for dye-sensitized solar cells are:

Parameters	Types of dye sensitized solar cells	
	Grätzel cell	Solid-state
Efficiency (%)	~ 10–11	~ 4
V_{oc} (V)	~ 0.7	~ 0.40
J_{sc} (mA/cm²)	~ 20	~ 9.10
Fill factor	~ 0.67	~ 0.6

Challenges

Liquid organic electrolytes contain highly corrosive iodine, leading to problems of leakage, sealing, handling, dye desorption, and maintenance. Much attention is now focused on the electrolyte to address these problems.

For solid-state dye sensitized solar cells, the first challenge originates from disordered titania mesoporous structures. Mesoporous titania structures should be fabricated with well-ordered titania structures of uniform size (~ 10 nm). The second challenge comes from developing the solid electrolyte, which is required to have these properties:

- The electrolyte should be transparent to the visible spectrum (wide band gap).

- Fabrication should be possible for depositing the solid electrolyte without degrading the dye molecule layer on titania.

- The LUMO of the dye molecule should be higher than the conduction band of titania.

- Several p-type semiconductors tend to crystallize inside the mesoporous titania films, destroying the dye molecule-titania contact. Therefore, the solid electrolyte needs to be stable during operation.

Nanostructured Inorganic — Small Molecules

In 2008, scientists were able to create a nanostructured lamellar structure that provides an ideal design for bulk heterojunction solar cells. The observed structure is composed of ZnO and small, conducting organic molecules, which co-assemble into alternating layers of organic and inorganic components. This highly organized structure, which is stabilized by π-π stacking between the organic molecules, allows for conducting pathways in both the organic and inorganic layers. The thicknesses of the layers (about 1–3 nm) are well within the exciton diffusion length, which ideally minimizes recombination among charge carriers. This structure also maximizes the interface between the inorganic ZnO and the organic molecules, which enables a high chromophore loading density within the structure. Due to the choice of materials, this system is non-toxic and environmentally friendly, unlike many other systems which use lead or cadmium.

Although this system has not yet been incorporated into a photovoltaic device, preliminary photoconductivity measurements have shown that this system exhibits among the highest values measured for organic, hybrid, and amorphous silicon photoconductors, and so, offers promise in creating efficient hybrid photovoltaic devices.

Photoredox Catalysis

Photoredox catalysis is a branch of catalysis that harnesses the energy of light to accelerate a chemical reaction via single-electron transfer events. This area is named as a combination of "photo-" referring to light and redox, a condensed expression for the chemical processes of reduction and oxidation. In particular, photoredox catalysis employs small quantities of a light-sensitive compound that, when excited by light, can mediate the transfer of electrons between chemical compounds that would usually not react at all. Photoredox catalysts are generally drawn from three classes of materials: transition-metal complexes, organic dyes, and semiconductors. While organic photoredox catalysts were dominant throughout the 1990s and early 2000s, soluble transition-metal complexes are more commonly used today.

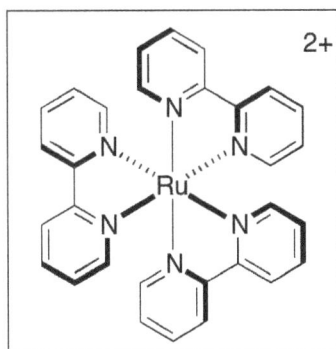

Study of this branch of catalysis led to the development of new methods to accomplish known and new chemical transformations. Photoredox catalysts are usually far less toxic than traditional reagents used to generate free radicals, such as organotin compounds. Furthermore, photoredox catalysts generate potent redox agents when exposed to light, they are unreactive under normal conditions. Thus, transition-metal complex photoredox catalysts are more attractive than stoichiometric redox agents such as quinones. The properties of transition-metal photoredox catalysts depend on the ligands and metal and can be modified for different purposes.

Photoredox catalysis is often applied to generate known reactive intermediates in a novel way and has led to the discovery of new organic reactions, such as the first direct functionalization of the β-arylation of saturated aldehydes. While the D_3-symmetric transition-metal complexes used in many photoredox-catalyzed reactions are chiral, enantioenriched photoredox catalysts has only led to low levels of enantioselectivity in a photoredox-catalyzed aryl-aryl coupling reaction, suggesting that the chiral nature of these catalysts is still poor at transmitting stereochemical information. While synthetically useful levels of enantioselectivity has not been achieved using chiral photoredox catalysts alone, enantioselectivity have been obtained through the synergistic combination of photoredox catalysis with chiral organocatalysts such as secondary amines and Brønsted acids.

Photochemistry of Transition Metal Sensitizers

Sensitizers absorb light to give redox-active excited states. For many metal-based sensitizers, excitation is realized as a metal-to-ligand charge transfer, whereby an electron moves from the metal (e.g., a d orbital) to an orbital localized on the ligands (e.g. the π* orbital of an aromatic ligand). The initial excited electronic state relaxes to the lowest energy singlet excited state through internal conversion, a process where energy is dissipated as vibrational energy rather than as electromagnetic radiation. This singlet excited state can relax further by two distinct processes: the catalyst may fluoresce, radiating a photon and returning to the singlet ground state, or it can move to the lowest energy triplet excited state (a state where two unpaired electrons have the same spin) by a second non-radiative process termed intersystem crossing.

Direct relaxation of the excited triplet to the ground state, termed phosphorescence, requires both emission of a photon and inversion of the spin of the excited electron. This pathway is slow because it is spin-forbidden so the triplet excited state has a substantial average lifetime. For the common photosensitizer, tris-(2,2'-bipyridyl)ruthenium (abbreviated as $[Ru(bipy)_3]^{2+}$ or $[Ru(bpy)_3]^{2+}$), the lifetime of the triplet excited state is approximately 1100 ns. This lifetime is sufficient for other relaxation pathways (specifically, electron-transfer pathways) to occur before decay of the catalyst to its ground state.

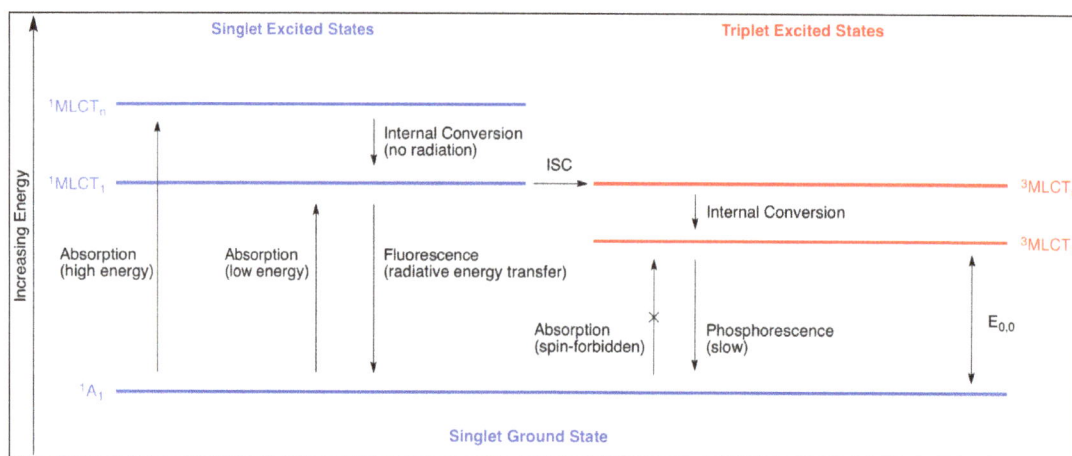

The long-lived triplet excited state accessible by photoexcitation is both a more potent reducing agent and a more potent oxidizing agent than the ground state of the catalyst. Since sensitizer is coordinatively saturated, electron transfer must occur by an outer sphere process, where the electron tunnels between the catalyst and the substrate.

Outer Sphere Electron Transfer

Marcus' theory of outer sphere electron transfer predicts that such a tunneling process will occur most quickly in systems where the electron transfer is thermodynamically favorable (i.e. between strong reductants and oxidants) and where the electron transfer has a low intrinsic barrier.

The intrinsic barrier of electron transfer derives from the Franck–Condon principle, stating that electronic transition takes place more quickly given greater overlap between the initial and final electronic states. Interpreted loosely, this principle suggests that the barrier of an electronic transition is related to the degree to which the system seeks to reorganize. For an electronic transition with a system, the barrier is related to the "overlap" between the initial and final wave functions of the excited electron–i.e. the degree to which the electron needs to "move" in the transition.

In an intermolecular electron transfer, a similar role is played by the degree to which the nuclei seek to move in response to the change in their new electronic environment. Immediately after electron transfer, the nuclear arrangement of the molecule, previously an equilibrium, now represents a vibrationally excited state and must relax to its new equilibrium geometry. Rigid systems, whose geometry is not greatly dependent on oxidation state, therefore experience less vibrational excitation during electron transfer, and have a lower intrinsic barrier. Photocatalysts such as $[Ru(bipy)_3]^{2+}$, are held in a rigid arrangement by flat, bidentate ligands arranged in an octahedral geometry around the metal center. Therefore, the complex does not undergo much reorganization during electron transfer. Since electron transfer of these complexes is fast, it is likely to take place within the duration of the catalyst's active state, i.e. during the lifetime of the triplet excited state.

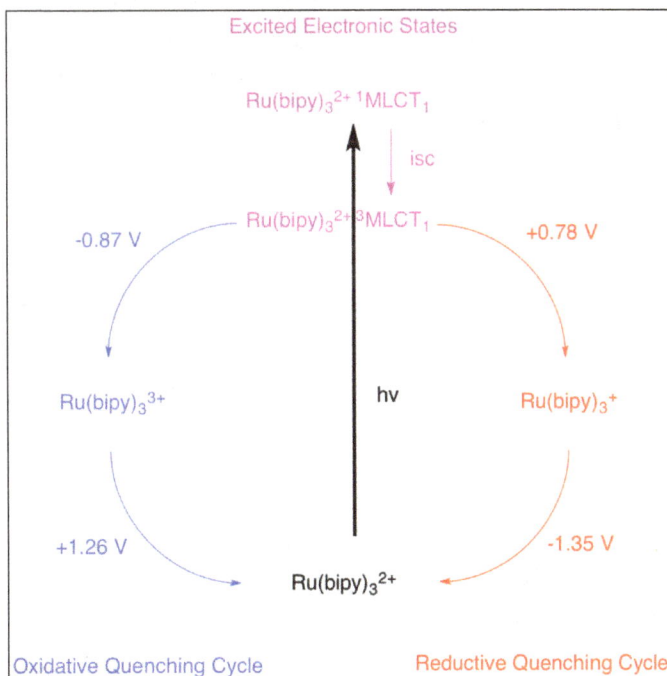

Catalyst Regeneration

The final step in the photocatalytic cycle is the regeneration of the photocatalyst in its ground state. At this stage, the catalyst exists as the ground state of either its oxidized or reduced forms, depending on whether it donated or accepted an electron. These oxidation states have a strong driving force to return to their equilibrium oxidation state and act as a potent single-electron reductant or oxidant to satisfy that driving force.

To regenerate the original ground state, the catalyst must participate in a second outer-sphere electron transfer. In many cases, this electron transfer takes place with a stoichiometric two-electron reductant or oxidant, although in some cases this step involves a second reagent. The reductive quenching cycle is when the excited state catalyst is first reduced then oxidized to return to its resting state. Conversely, the oxidative quenching cycle is when the excited state catalyst is first oxidized and then reduced to return to its resting state. These two cycles can be distinguished by a Stern–Volmer experiment.

Since the electron transfer step of the catalytic cycle takes place from the triplet excited state, it competes with phosphorescence as a relaxation pathway. The Stern–Volmer experiment measures the intensity of phosphorescence while varying the concentration of each possible quenching agent. When the concentration of the actual quenching agent is varied, the rate of electron transfer and the degree of phosphorescence is affected. This relationship is modeled by the equation:

$$\left(\frac{I_0}{I}\right) = 1 + k_q * \tau_0 \times [Q]$$

Here, I_0 and I denote the emission intensity with and without quenching agent present, k_q the rate constant of the quenching process, τ_0 the excited-state lifetime in the absence of quenching agent and $[Q]$ the concentration of quenching agent. Thus, if the excited-state lifetime of the photoredox catalyst is known from other experiments, the rate constant of quenching in the presence of a single reaction component can be determined by measuring the change in emission intensity as the concentration of quenching agent changes.

Photophysical Properties

Redox Potentials

The redox potentials of photoredox catalysts must be matched to the reaction's other components. While ground state redox potentials are easily measured by cyclic voltammetry or other electrochemical methods, measuring the redox potential of an electronically excited state cannot be accomplished directly by these methods. However, two methods exist that allow estimation of the excited-state redox potentials and one method exists for the direct measurement of these potentials. To estimate the excited-state redox potentials, one method is to compare the rates of electron transfer from the excited state to a series of ground-state reactants whose redox potentials are known. A more common method to estimate these potentials is to use an equation developed by Rehm and Weller that describes the excited-state potentials as a correction of the ground-state potentials:

$$E^*_{1/2}{}^{red} = E_{1/2}{}^{red} + E_{0,0} + w_r$$

$$E^*_{1/2}{}^{ox} = E_{1/2}{}^{ox} - E_{0,0} + w_r$$

In these formulas, $E^*_{1/2}$ represents the reduction or oxidation potential of the excited state, $E_{1/2}$ represents the reduction or oxidation potential of the ground state, $E_{0,0}$ represents the difference in energy between the zeroth vibrational states of the ground and excited states and w_r represents the work function, an electrostatic interaction that arises due to the separation of charges that occurs during electron-transfer between two chemical species. The zero-zero excitation energy, $E_{0,0}$ is usually approximated by the corresponding transition in the fluorescence spectrum. This method allows calculation of approximate excited-state redox potentials from more easily measured ground-state redox potentials and spectroscopic data.

Direct measurement of the excited-state redox potentials is possible by applying a method known as phase-modulated voltammetry. This method works by shining light onto an electrochemical cell in order to generate the desired excited-state species, but to modulate the intensity of the light sinusoidally, so that the concentration of the excited-state species is not constant. In fact, the concentration of excited-state species in the cell should change exactly in phase with the intensity of light incident on the electrochemical cell. If the potential applied to the cell is strong enough for electron transfer to occur, the change in concentration of the redox-competent excited state can be measured as an alternating current (AC). Furthermore, the phase shift of the AC current relative to the intensity of the incident light corresponds to the average lifetime of an excited-state species before it engages in electron transfer.

Charts of redox potentials for the most common photoredox catalysts are available for quick access.

Ligand Electronegativity

The relative reducing and oxidizing natures of these photocatalysts can be understood by considering the ligands' electronegativity and the catalyst complex's metal center. More electronegative metals and ligands can stabilize electrons better than their less electronegative counterparts. Therefore, complexes with more electronegative ligands are more oxidising than less electronegative ligand complexes. For example, the ligands 2,2'-bipyridine and 2,2'-phenylpyridine are isoelectronic structures, containing the same number and arrangement of electrons. Phenylpyridine replaces one of the nitrogen atoms in bipyridine with a carbon atom. Carbon is less electronegative than nitrogen is, so it holds electrons less tightly. Since the remainder of the ligand molecule is identical and phenylpyridine holds electrons less tightly than bipyridine, it is more strongly electron-donating and less electronegative as a ligand. Hence, complexes with phenylpyridine ligands are more strongly reducing and less strongly oxidizing than equivalent complexes with bipyridine ligands.

Similarly, a fluorinated phenylpyridine ligand is more electronegative than phenylpyridine so complexes with fluorine-containing ligands are more strongly oxidizing and less strongly reducing than equivalent unsubstituted phenylpyridine complexes. The metal center's electronic influence on the complex is more complex than the ligand effect. According to the Pauling scale of electronegativity, both ruthenium and iridium have an electronegativity of 2.2. If this was the sole factor relevant to redox potentials, then complexes of ruthenium and iridium with the same ligands should be equally powerful photoredox catalysts. However, considering the Rehm-Weller equation, the spectroscopic properties of the metal play a role in determining the redox properties of the excited state. In particular, the parameter $E_{0,0}$ is related to the emission wavelength of the complex and therefore, to the size of the Stokes shift - the difference in energy between the maximum absorption and emission of a molecule. Typically, ruthenium complexes have large Stokes shifts and hence, low energy emission wavelengths and small zero-zero excitation energies when compared to iridium complexes. In effect, while ground-state ruthenium complexes can be potent reductants, the excited-state complex is a far less potent reductant or oxidant than its equivalent iridium complex. This makes iridium preferred for the development of general organic transformations because the stronger redox potentials of the excited catalyst allows the use of weaker stoichiometric reductants and oxidants or the use of less reactive substrates.

Applications

Reductive Dehalogenation

Reductive dehalogenation is the removing of halogen atoms from a molecule. However, the traditional method for dehalogenation uses stoichiometric organotin reagents, such as tributyltin hydride. While this reaction is powerful with high functional group tolerance, organotin reagents are highly toxic. The cleavage of activated and reductively labile functional groups including sulfoniums and halogens is the earliest application of photoredox catalysis to organic synthesis, but early attempts were limited by the need for specific substrates or by the formation of dimeric coupling products. More general methods are known. One method employs [Ru(bipy)$_3$]$^{2+}$ as the photocatalyst and a stoichiometric amine reductant to reduce "activated" carbon-halogen bonds, such as those with an adjacent carbonyl group or arene. These bonds are considered to be activated because the radical they produce upon fragmentation is stabilized by conjugation with the carbonyl group or arene, respectively. The stoichiometric reductant present in this reaction transfers an electron to reduce the excited-state catalyst to the Ru(I) oxidation state. The reduced catalyst then shuttles the transferred electron to the halogenated substrate, reducing the weak C-X bond and inducing fragmentation.

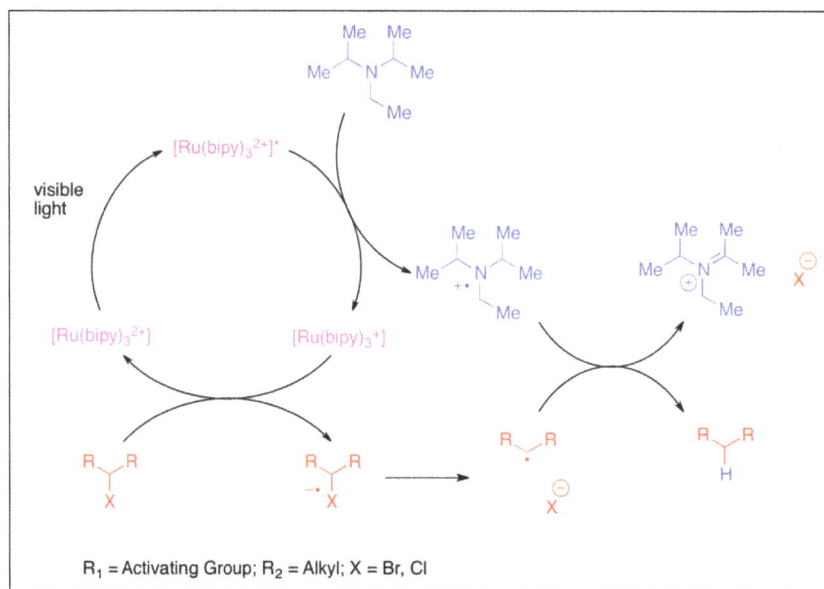

R_1 = Activating Group; R_2 = Alkyl; X = Br, Cl

Unactivated carbon-iodine bonds can be reduced using the strongly reducing photocatalyst tris-(2,2'-phenylpyridine)iridium (Ir(ppy)$_3$). This reaction is mechanistically distinct from the previous transformation of activated bromides and chlorides. The increased reduction potential of Ir(ppy)$_3$ compared to [Ru(bipy)$_3$]$^{2+}$ allows direct reduction of the carbon-iodine bond without interacting with a stoichiometric reductant. Thus, the iridium complex transfers an electron to the substrate, causing fragmentation of the substrate and oxidizing the catalyst to the Ir(IV) oxidation state. The oxidized photocatalyst is returned to its original oxidation state by oxidising a reaction additives.

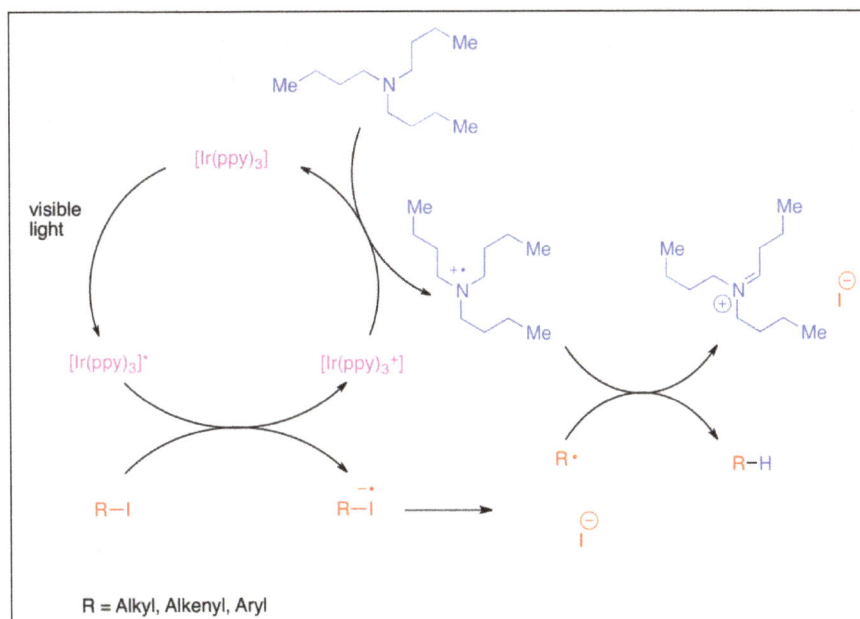

R = Alkyl, Alkenyl, Aryl

Like tin-mediated radical dehalogenation reactions, photocatlytic reductive dehalogenation can be used to initiate cascade cyclizations to rapidly generate molecular complexity. In this work, a radical cascade cyclization that closed two five-membered rings and formed two new stereocenters, in good yield. This reductive dehalogenation protocol was a key step in a total synthesis of the natural product (+)-Gliocladin C.

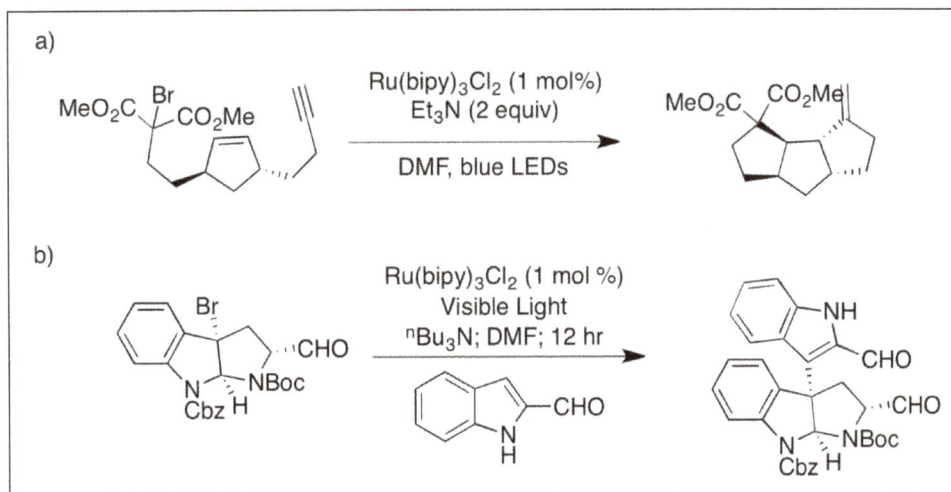

Oxidative Generation of Iminium Ions

Iminium ions are potent electrophiles useful for generating C-N bonds in complex molecules. However, the condensation of amines with carbonyl compounds to form iminium ions is often unfavorable, sometimes requiring harsh dehydrating conditions. Thus, alternative methods for iminium ion generation, particularly by oxidation from the corresponding amine, are a valuable synthesis tool. Iminium ions can be generated from activated amines using Ir(dtbbpy)(ppy)$_2$PF$_6$ as a photoredox catalyst. This transformation is proposed to occur by oxidation of the amine to the aminium radical cation by the excited photocatalyst. This is followed by hydrogen atom transfer to a superstoichimetric oxidant, such as trichloromethyl radical (CCl$_3$ to form the iminium ion). The iminium ion is then quenched by reaction with a nucleophile. Related transformations of amines with a wide variety of other nucleophiles have been investigated, such as cyanide (Strecker reaction), silyl enol ethers (Mannich reaction), dialkylphosphates, allyl silanes (aza-Sakurai reaction), indoles (Friedel-Crafts reaction), and copper acetylides.

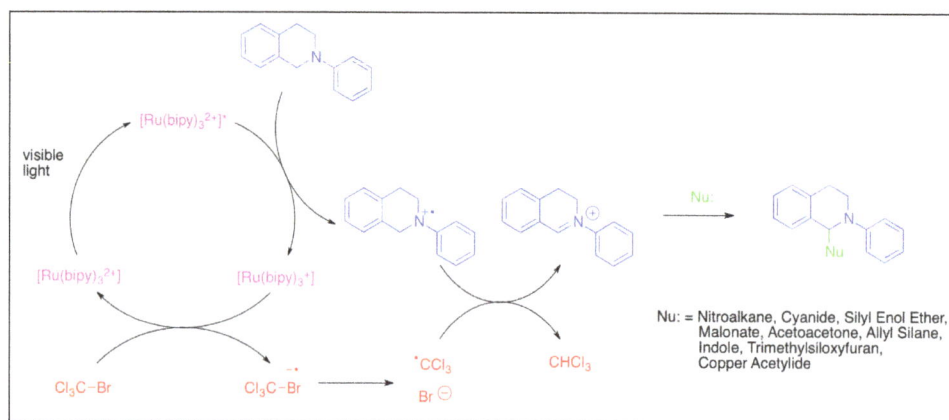

Similar photoredox generation of iminium ions has furthermore been achieved using purely organic photoredox catalysts, such as Rose Bengal and Eosin Y.

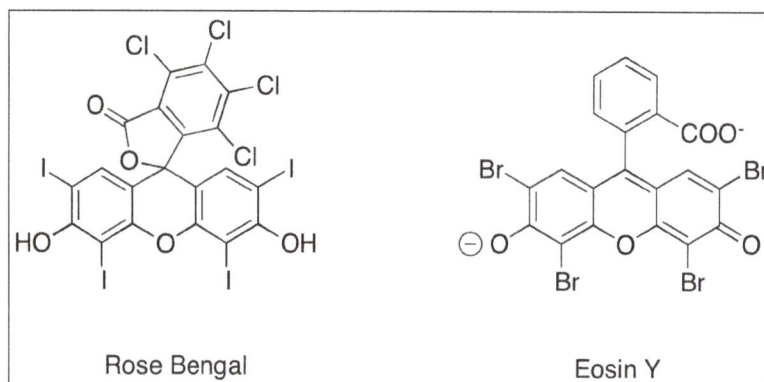

Rose Bengal

Eosin Y

An asymmetric variant of this reaction utilizes acyl nucleophile equivalents generated by N-heterocyclic carbene catalysis. This reaction method sidesteps the problem of poor enantioinduction from chiral photoredox catalysts by moving the source of enantioselectivity to the N-heterocyclic carbene.

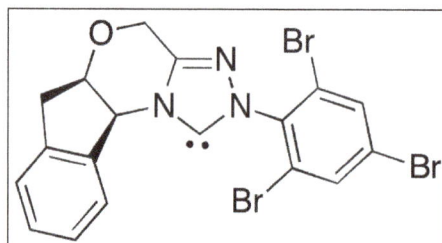

Oxidative Generation of Oxocarbenium Ions

The development of orthogonal protecting groups is a problem in organic synthesis because these protecting groups allow each instance of a common functional group, such as the hydroxyl group, to be distinguished during the synthesis of a complex molecule. A very common protecting group for the hydroxyl functional group is the *para*-methoxy benzyl (PMB) ether. This protecting group is chemically similar to the less electron-rich benzyl ether. Typically, selective cleavage of a PMB ether in the presence of a benzyl ether uses strong stoichiometric oxidants such as 2,3-dichloro-5,6-dicyano-1,4-benzoquinone (DDQ) or ceric ammonium nitrate (CAN). PMB ethers are far more susceptible to oxidation than benzyl ethers since they are more electron-rich. The selective deprotection of PMB ethers can be achieved through the use of bis-(2-(2',4'-difluorophenyl)-5-trifluoromethylpyridine)-(4,4'-ditertbutylbipyridine)iridium(III) hexafluorophosphate (Ir[dF(CF$_3$)ppy]$_2$(dtbbpy)PF$_6$) and a mild stoichiometric oxidant such as bromotrichloromethane, BrCCl$_3$. The photoexcited iridium catalyst is reducing enough to fragment the bromotrichloromethane to form a trichloromethyl radical, bromide anion, and the Ir(IV) complex. The electron-poor fluorinated ligands makes the iridium complex oxidising enough to accept an electron from an electron-rich arene such as a PMB ether. After the arene is oxidized, it will readily participate in hydrogen atom transfer with trichloromethyl radical to form chloroform and an oxocarbenium ion, which is readily hydrolyzed to reveal the free hydroxide. This reaction was demonstrated to be orthogonal to many common protecting groups when a base was added to neutralise the HBr produced.

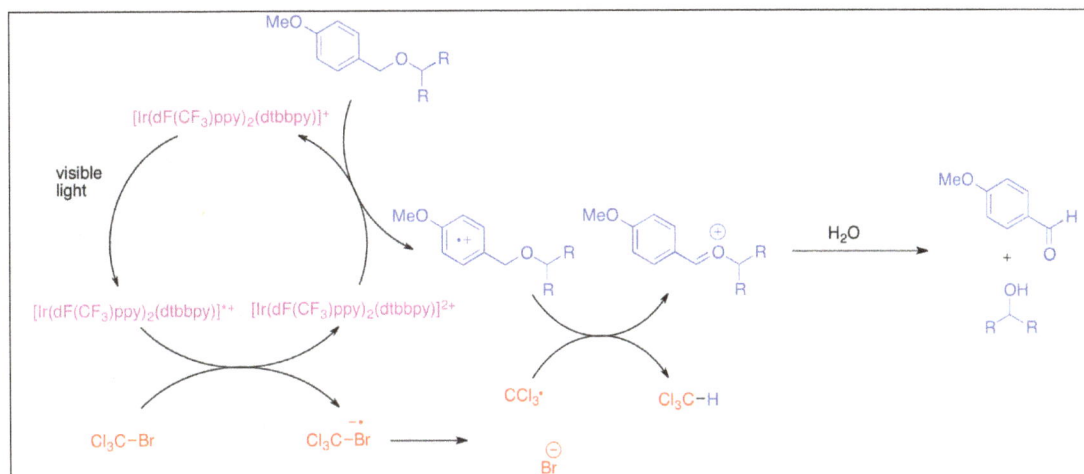

Cycloadditions

Cycloadditions and other pericyclic reactions are powerful transforms in organic synthesis because of their potential to rapidly generate complex molecular architectures and particularly because of their capacity to set multiple adjacent stereocenters in a highly controlled manner. However, only certain cycloadditions are allowed under thermal conditions according to the Woodward–Hoffmann rules of orbital symmetry, or other equivalent models such as frontier molecular orbital theory (FMO) or the Dewar-Zimmermann model. Cycloadditions that are not thermally allowed, such as the [2+2] cycloaddition, can be enabled by photochemical activation of the reaction. Under uncatalyzed conditions, this activation requires the use of high energy ultraviolet light capable of altering the orbital populations of the reactive compounds. Alternatively, metal catalysts such as cobalt and copper have been reported to catalyze thermally-forbidden [2+2] cycloadditions via single electron transfer.

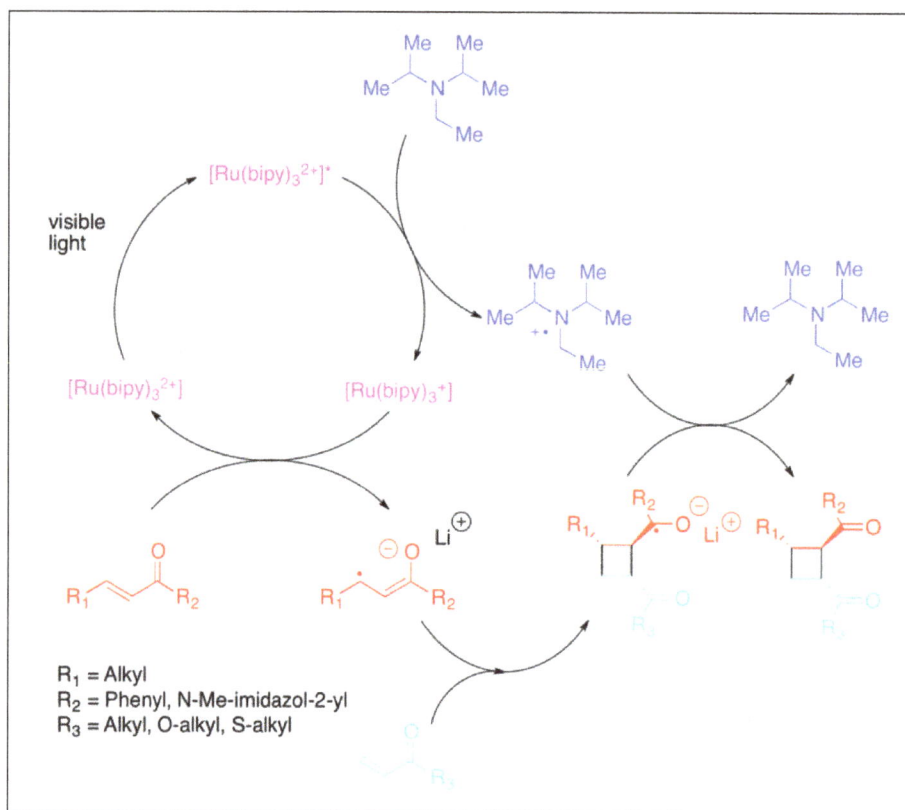

R₁ = Alkyl
R₂ = Phenyl, N-Me-imidazol-2-yl
R₃ = Alkyl, O-alkyl, S-alkyl

The required change in orbital populations can be achieved by electron transfer with a photocatalyst sensitive to lower energy visible light. Yoon demonstrated the efficient intra- and intermolecular [2+2] cycloadditions of activated olefins: particularly enones and styrenes. Enones, or electron-poor olefins, were discovered to react via a radical-anion pathway, utilizing diisopropylethylamine as a transient source of electrons. For this electron-transfer, [Ru(bipy)$_3$]$^{2+}$ was discovered to be an efficient photocatalyst. The anionic nature of the cyclization proved to be crucial: performing the reaction in acid rather than with a lithium counterion favored a non-cycloaddition pathway. Zhao et al. likewise discovered that a still different cyclization pathway is available to chalcones with a samarium counterion. Conversely, electron-rich styrenes were found to react via a radical-cation mechanism, utilizing methyl viologen or molecular oxygen as a transient electron sink. While [Ru(bipy)$_3$]$^{2+}$ proved to be a competent catalyst for intramolecular cyclizations using methyl viologen, it could not be used with molecular oxygen as an electron sink or for intermolecular cyclizations. For intermolecular cyclizations, Yoon et al. discovered that the more strongly oxidizing photocatalyst [Ru(bpm)$_3$]$^{2+}$ and molecular oxygen provided a catalytic system better suited to access the radical cation necessary for the cycloaddition to occur. [Ru(bpz)$_3$]$^{2+}$, a still more strongly oxidizing photocatalyst, proved to be problematic because although it could catalyze the desired [2+2] cycloaddition, it was also strong enough to oxidize the cycloadduct and catalyze the retro-[2+2] reaction. This comparison of photocatalysts highlights the importance of tuning the redox properties of a photocatalyst to the reaction system as well as demonstrating the value of polypyridyl compounds as ligands, due to the ease with which they can be modified to adjust the redox properties of their complexes.

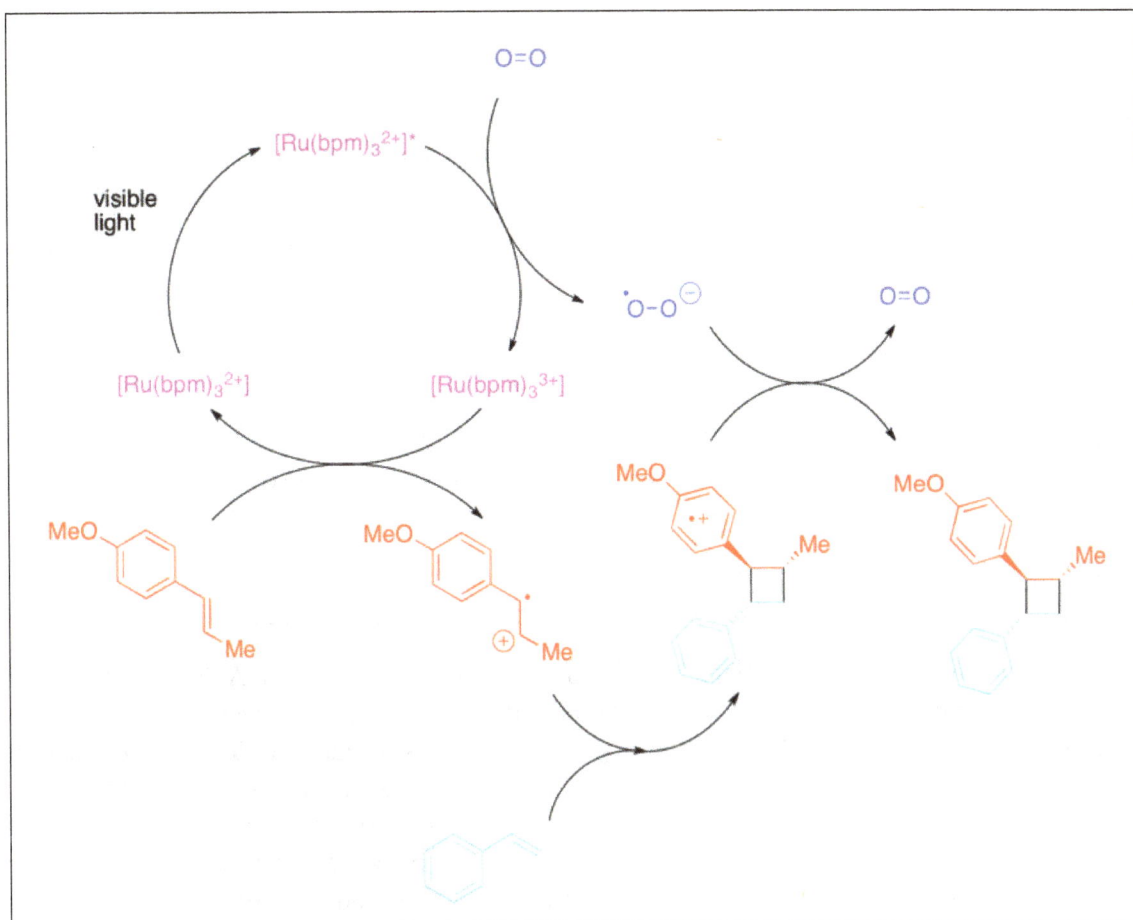

Photoredox-catalyzed [2+2] cycloadditions can also be effected with a triphenylpyrylium organic photoredox catalyst.

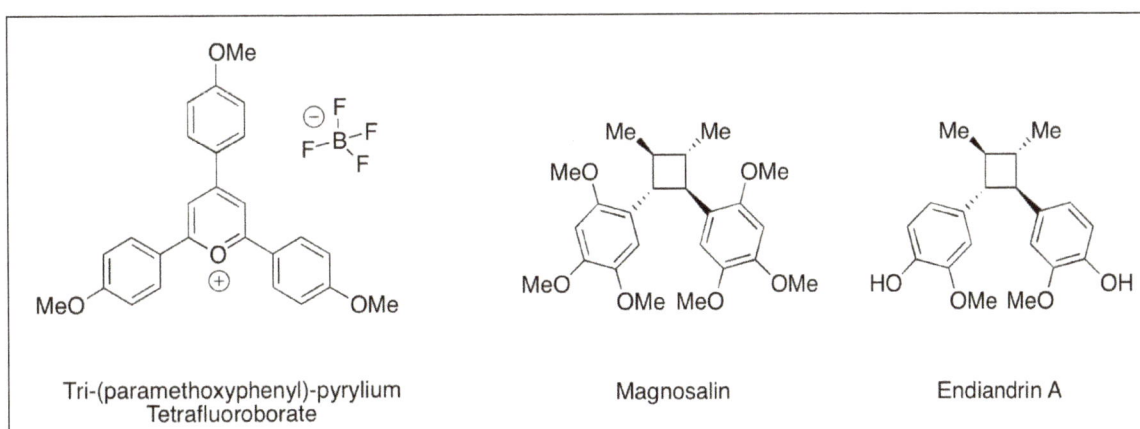

Tri-(paramethoxyphenyl)-pyrylium
Tetrafluoroborate

Magnosalin

Endiandrin A

In addition to the thermally-forbidden [2+2] cycloaddition, photoredox catalysis can be applied to the [4+2] cyclization (Diels–Alder reaction). Bis-enones, similar to the substrates used for the photoredox [2+2] cyclization, but with a longer linker joining the two enone functional groups, undergo intramolecular radical-anion hetero-Diels–Alder reactions more rapidly than [2+2] cycloaddition.

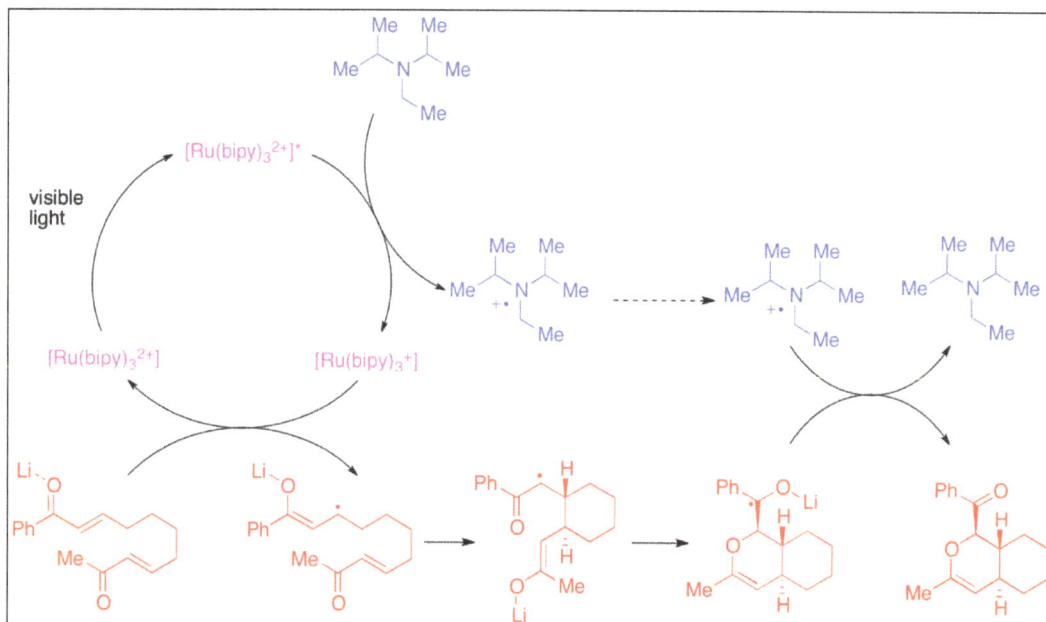

Similarly, electron-rich styrenes participate in intra- or intermolecular Diels–Alder cyclizations via a radical cation mechanism. [Ru(bipy)$_3$]$^{2+}$ was a competent catalyst for intermolecular, but not intramolecular, Diels–Alder cyclizations. This photoredox-catalyzed Diels–Alder reaction allows cycloaddition between two electronically mismatched substrates. The normal electronic demand for the Diels–Alder reaction calls for an electron-rich diene to react with an electron-poor olefin (or "dienophile"), while the inverse electron-demand Diels–Alder reaction takes place between the opposite case of an electron-poor diene and a very electron-rich dienophile. The photoredox case, since it takes place by a different mechanism than the thermal Diels–Alder reaction, allows cycloaddition between an electron-rich diene and an electron-rich dienophile, allowing access to new classes of Diels–Alder adducts.

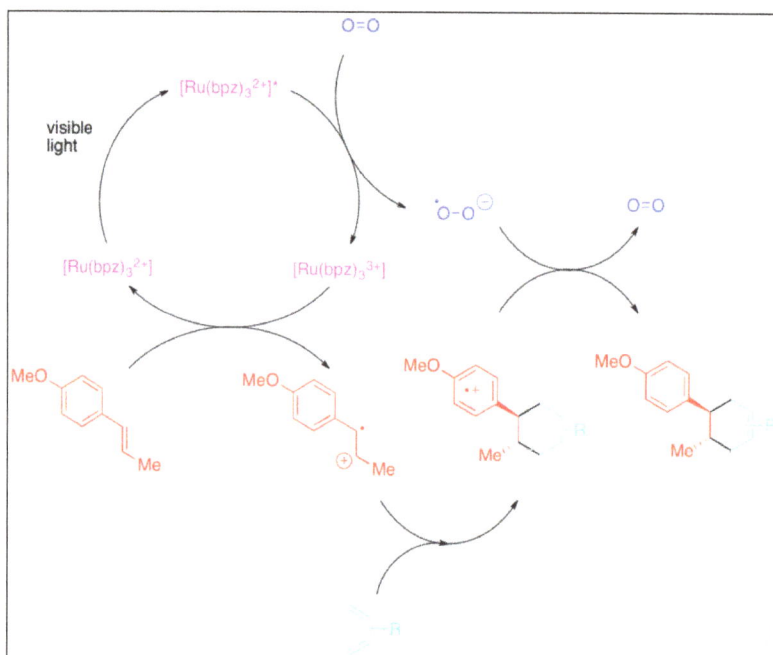

The synthetic value of Yoon's photoredox-catalyzed styrene Diels–Alder reaction was demonstrated via the total synthesis of the natural product Heitziamide A. This synthesis demonstrates that the thermal Diels–Alder reaction favors the undesired regioisomer, but the photoredox-catalyzed reaction gives the desired regioisomer in improved yield.

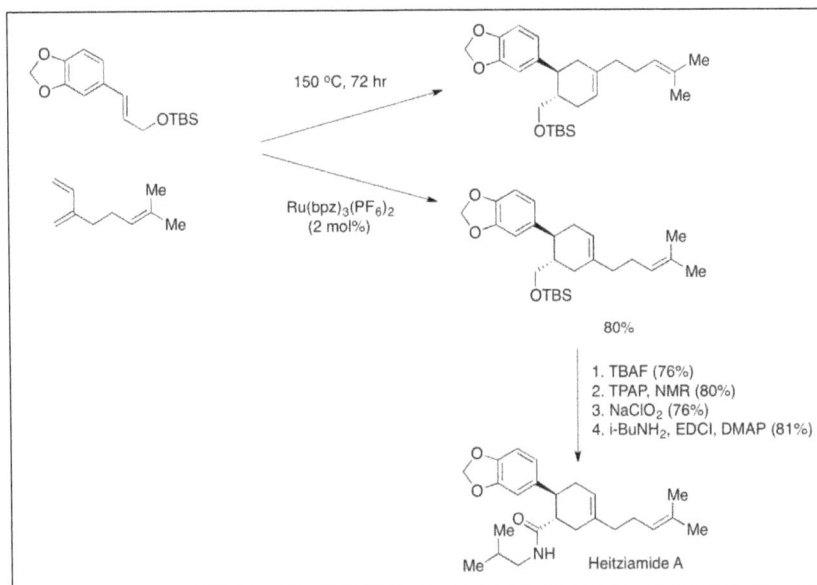

Photoredox Organocatalysis

Organocatalysis is a subfield of catalysis that explores the potential of organic small molecules as catalysts, particularly for the enantioselective creation of chiral molecules. One strategy in this subfield is the use of chiral secondary amines to activate carbonyl compounds. In this case, amine condensation with the carbonyl compound generates a nucleophilic enamine. The chiral amine is designed so that one face of the enamine is sterically shielded and so that only the unshielded face is free to react. Despite the power of this approach to catalyze the enantioselective functionalization of carbonyl compounds, certain valuable transformations, such as the catalytic enantioselective α-alkylation of aldehydes, remained elusive. The combination of organocatalysis and photoredox methods provides a catalytic solution to this problem. In this approach for the α-alkylation of aldehydes, $[Ru(bipy)_3]^{2+}$ reductively fragments an activated alkyl halide, such as bromomalonate or phenacyl bromide, which can then add to catalytically-generated enamine in an enantioselective manner. The oxidized photocatalyst then oxidatively quenches the resulting α-amino radical to form an iminium ion, which hydrolyzes to give the functionalized carbonyl compound. This photoredox transformation was shown to be mechanistically distinct from another organocatalytic radical process termed singly-occupied molecular orbital (SOMO) catalysis. SOMO catalysis employs superstoichiometric ceric ammonium nitrate (CAN) to oxidize the catalytically-generated enamine to the corresponding radical cation, which can then add to a suitable coupling partner such as allyl silane. This type of mechanism is excluded for the photocatalytic alkylation reaction because whereas enamine radical cation was observed to cyclize onto pendant olefins and open cyclopropane radical clocks in SOMO catalysis, these structures were unreactive in the photoredox reaction.

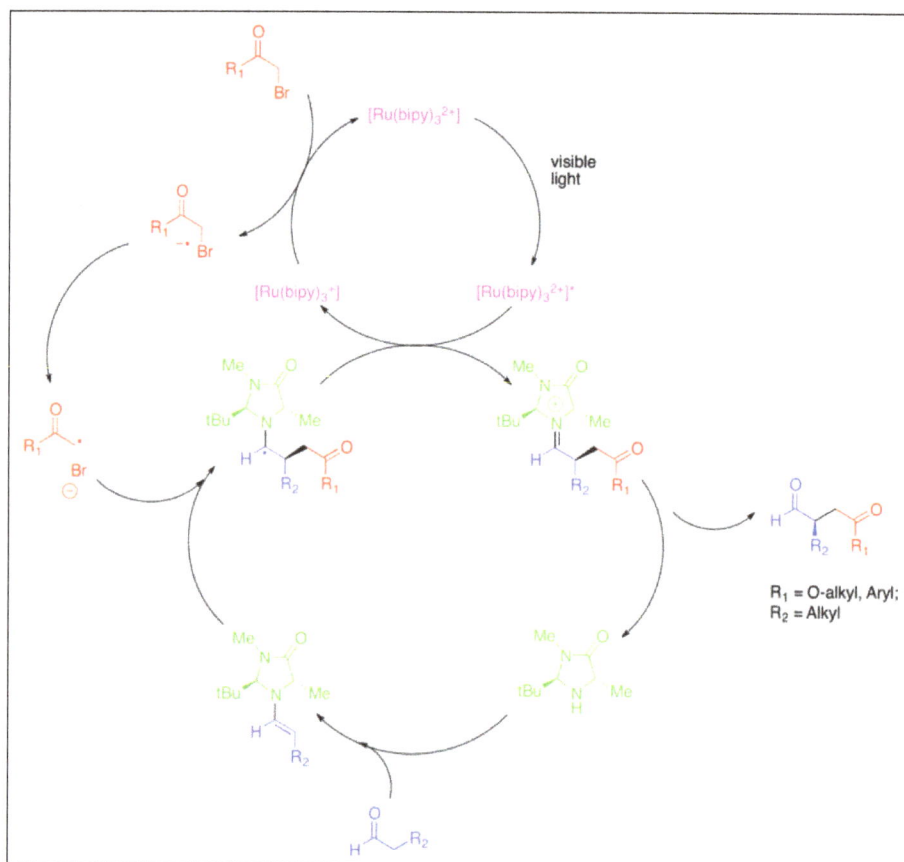

This transformation include alkylations with other classes of activated alkyl halides of synthetic interest. In particular, the use of the photocatalyst Ir(dtbbpy)(ppy)$_2{}^+$ allows the enantioselective α-trifluoromethylation of aldehydes while the use of Ir(ppy)$_3$ allowed the enantioselective coupling of aldehydes with electron-poor benzylic bromides. Zeitler et al. also investigated the productive merger of photoredox and organocatalytic methods to achieve enantioselective alkylation of aldehydes. The same chiral imidazolidinone organocatalyst was used to form enamine and introduce chirality. However, the organic photoredox catalyst Eosin Y was used rather than a ruthenium or iridium complex.

Direct β-arylation of saturated aldehydes and ketones can be effected through the combination of photoredox and organocatalytic methods. The previous method to accomplish direct β-functionalization of a saturated carbonyl consists of a one-pot consists of a two-step process, both catalyzed by a secondary amine organocatalyst: stoichiometric reduction of an aldehyde with IBX followed by addition of an activated alkyl nucleophile to the beta-position of the resulting enal. This transformation, which like other photoredox processes takes place by a radical mechanism, is limited to the addition of highly electrophilic arenes to the beta position. The severe limitations on the arene component scope in this reaction is due primarily to the need for an arene radical anion that is stable enough not to react directly with enamine or enamine radical cation. In the proposed mechanism, the activated photoredox catalyst is quenched oxidatively by an electron-deficient arene, such as 1,4-dicyanobenzene. The photocatalyst then oxidizes an enamine species, transiently generated by the condensation of an aldehyde with a secondary amine cocatalyst, such as the optimal isopropyl benzylamine. The resulting enamine radical cation usually reacts as a 3 π-electron system, but due to the stability of the radical coupling partners, deprotonation of the β-methylene position gives rise to a 5 π-electron system with strong radical character at the newly accessed β-carbon. Although this reaction relies on the use of a secondary amine organocatalyst to generate the enamine species which is oxidized in the proposed mechanism, no enantioselective variant of this reaction exists.

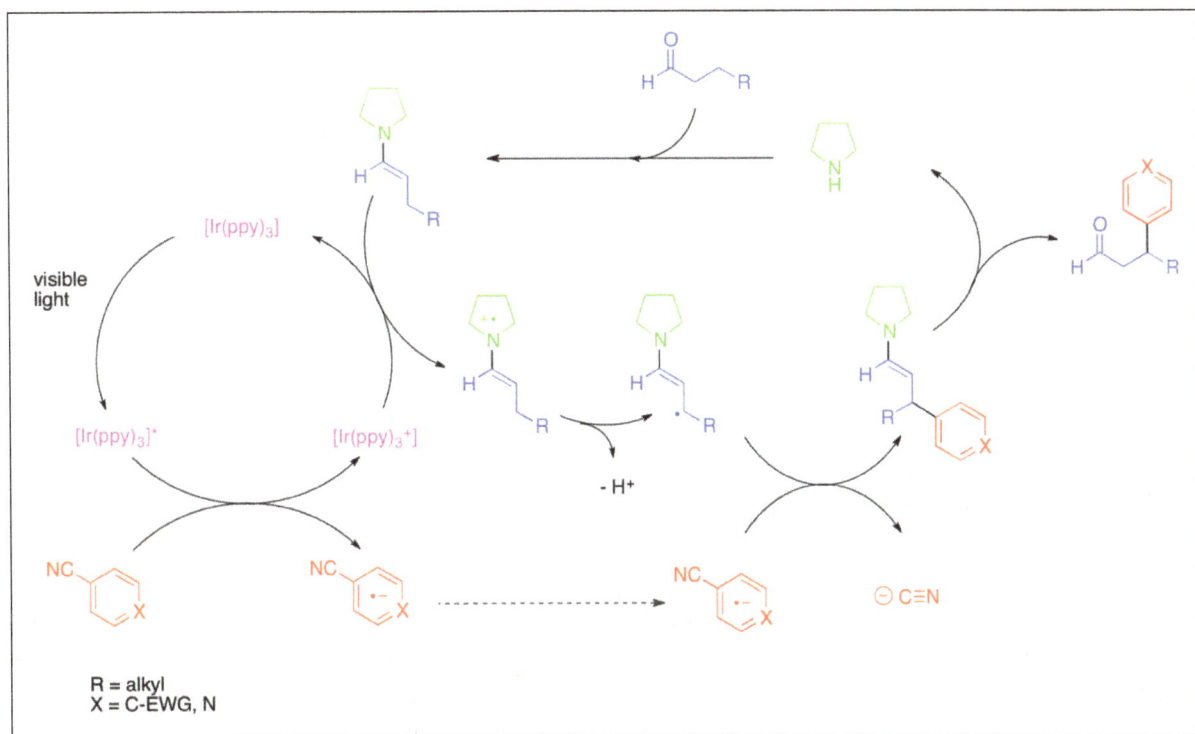

The development of this direct β-arylation of aldehydes led to related reactions for the β-functionalization of cyclic ketones. In particular, β-arylation of cyclic ketones has been achieved under similar reaction conditions, but using azepane as the secondary amine cocatalyst. A photocatalytic "homo-aldol" reaction works for cyclic ketones, allowing the coupling of the beta-position of the ketone to the ipso carbon of aryl ketones, such as benzophenone and acetophenone. In addition to the azepane cocatalyst, this reaction requires the use of the more strongly reducing photoredox catalyst Ir(ppy)$_3$ and the addition of lithium hexafluoroarsenide (LiAsF$_6$) to promote single-electron reduction of the aryl ketone.

Additions to Olefins

The use of photoredox catalysis to generate reactive heteroatom-centered radicals was first explored in the 1990s. [Ru(bipy)$_3$]$^{2+}$ was found to catalyze the fragmentation of tosylphenylselenide to phenylselenolate anion and tosyl radical and that a radical chain propagation mechanism allowed the addition of tosyl radical and phenylseleno- radical across the double bond of electron rich alkyl vinyl ethers. Since phenylselenolate anion is readily oxidized to diphenyldiselenide, the low quantities of diphenyldiselenide observed was taken as an indication that photoredox-catalyzed fragmentation of tosylphenylselenide was only important as an initiation step, and that most of the reactivity was due to a radical chain process.

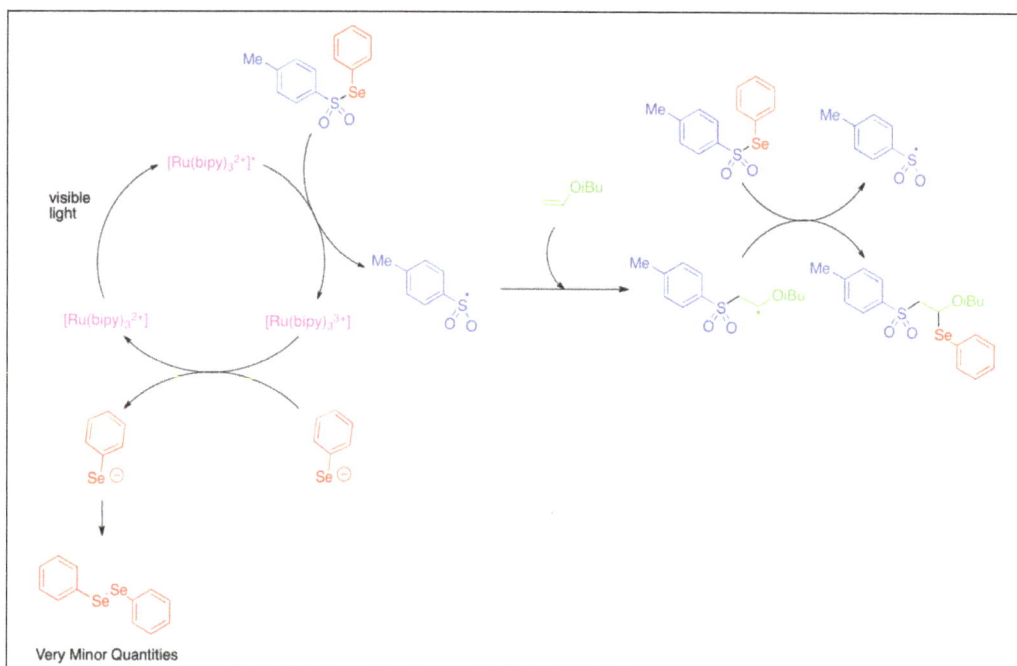

Very Minor Quantities

Heteroaromatic additions to olefins include multicomponent oxy- and aminotrifluoromethylation reactions. These reactions use Umemoto's reagent, a sulfonium salt that serves as an electrophilic source of the trifluoromethyl group and that is precedented to react via a single-electron transfer pathway. Thus, single-electron reduction of Umemoto's reagent releases trifluoromethyl radical, which adds to the reactive olefin. Subsequently, single-electron oxidation of the alkyl radical generated by this addition produces a cation which can be trapped by water, an alcohol, or a nitrile. In order to achieve high levels of regioselectivity, this reactivity has been explored mainly for styrenes, which are biased towards formation of the benzylic radical intermediate.

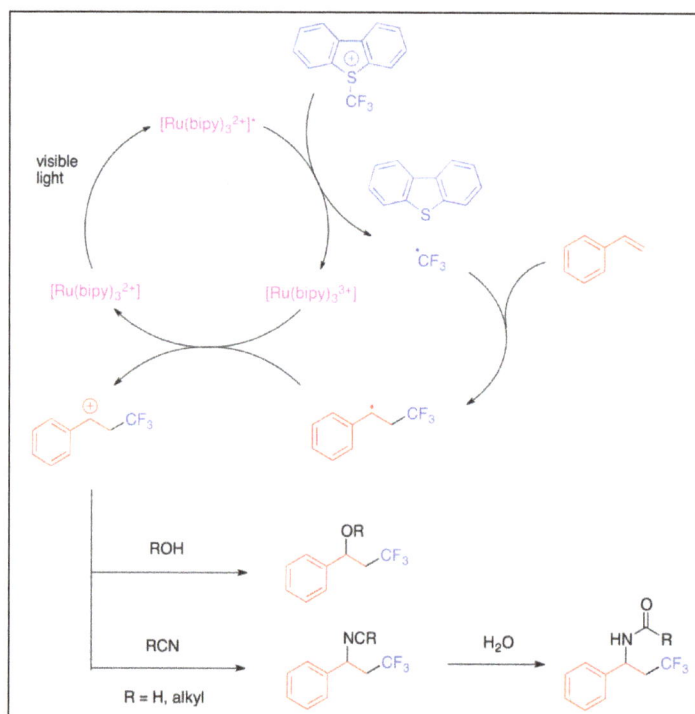

Hydrotrifluoromethylation of styrenes and aliphatic alkenes can be effected with a mesityl acridinium organic photoredox catalyst and Langlois' reagent as the source of CF_3 radical. In this reaction, it was found that trifluoroethanol and substoichiometric amounts of an aromatic thiol, such as methyl thiosalicylate, employed in tandem served as the best source of hydrogen radical to complete the catalytic cycle.

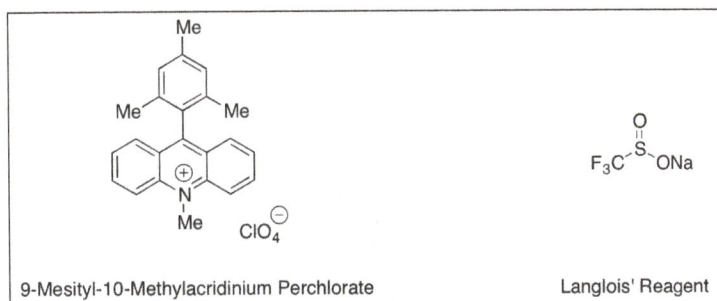

9-Mesityl-10-Methylacridinium Perchlorate   Langlois' Reagent

Intramolecular hydroetherifications and hydroaminations proceed with anti-Markovnikov selectivity. One mechanism invokes the single-electron oxidation of the olefin, trapping the radical cation by a pendant hydroxyl or amine functional group, and quenching the resulting alkyl radical by H-atom transfer from a highly labile donor species. Extensions of this reactivity to intermolecular systems have resulted in i) a new synthetic route to complex tetrahydrofurans by a "polar-radical-crossover cycloaddition" (PRCC reaction) of an allylic alcohol with an olefin, and ii) the anti-Markovnikov addition of carboxylic acids to olefins.

Photoelectric Effect

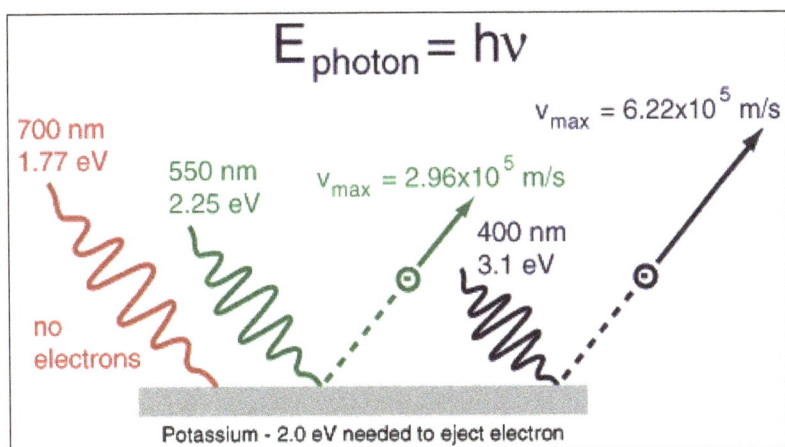

$$E_{photon} = h\nu$$

Low frequency light (red) is unable to cause ejection of electrons from the metal surface. At or above the threshold frequency (green) electrons are ejected. Even higher frequency incoming light (blue) causes ejection of the same number of electrons but with greater speed.

The photoelectric effect is a phenomenon that occurs when light shined onto a metal surface causes the ejection of electrons from that metal. It was observed that only certain frequencies of light are able to cause the ejection of electrons. If the frequency of the incident light is too low (red light, for example), then no electrons were ejected even if the intensity of the light was very high or it was shone onto the surface for a long time. If the frequency of the light was higher (green light, for example), then electrons were able to be ejected from the metal surface even if the intensity of the light was very low or it was shone for only a short time. This minimum frequency needed to cause electron ejection is referred to as the threshold frequency.

Classical physics was unable to explain the photoelectric effect. If classical physics applied to this situation, the electron in the metal could eventually collect enough energy to be ejected from the surface even if the incoming light was of low frequency. Einstein used the particle theory of light to explain the photoelectric effect as shown in the figure.

Photoelectric cells convert light energy into electrical energy which powers this calculator.

Consider the $E = h\upsilon$ equation. The E is the minimum energy that is required in order for the metal's electron to be ejected. If the incoming light's frequency, υ, is below the threshold frequency, there will never be enough energy to cause electron to be ejected. If the frequency is equal to or higher than the threshold frequency, electrons will be ejected.

As the frequency increases beyond the threshold, the ejected electrons simply move faster. An increase in the intensity of incoming light that is above the threshold frequency causes the number of electrons that are ejected to increase, but they do not travel any faster. The photoelectric effect is applied in devices called photoelectric cells, which are commonly found in everyday items such as a calculator which uses the energy of light to generate electricity.

The photoelectric effect was key to the development of the idea of a "photon" or the relationship of the energy of light to its frequency. The photoelectric effect is simply the effect that sometimes when you shine light on a metal, electrons are ejected. There are several key findings that we investigated in class.

- Unless light of sufficient frequency is used, then no electrons are ejected. That is there is a threshold below which no matter how intense the light source is, no electrons leave the metal.

- If you are using light of a sufficient frequency, then as the light source is increased in intensity (brightness), the number of electrons ejected increases.

- As the frequency is increased above the threshold, the velocity of the ejected electrons increases.

From this we can conclude that energy is proportional to frequency and that the proportionality constant is Planck's constant.

$$E = h\nu$$

Planck's Constant, $h = 6.626 \times 10^{-34} \, J \, s$.

We can also examine the relationship between the kinetic energy (E_k) of the electron and the frequency of the light used in the experiment. The maximum kinetic energy of the electron is the energy of the photon minus the threshold energy. This threshold energy we call the "work function" and we give it the symbol Φ.

$$E_K = \frac{1}{2}mv^2 = h\nu = \Phi$$

So we can either predict the maximum velocity of the electron for a given frequency if we know the work function or by measuring the maximum velocity for a given frequency we can calculate the work function.

Another way to think about this is simply conservation of energy. The energy of the photon must equal the sum of the work function (the potential energy that needs to be overcome for the electron to "escape") plus the kinetic energy of the electron. This is the same as the above equation re-arranged.

$$h\nu = E_K + \Phi = \frac{1}{2}mv^2 + \Phi$$

Emission Mechanism

All atoms have their electrons in orbitals with well-defined energy levels. When electromagnetic radiation interacts with an atom, it can excite the electron to a higher energy level, which can then fall back down, returning to the ground state. However, if the energy of the light is such that the electron is excited above energy levels associated with the atom, the electron can actually break free from the atom leading to ionization of the atom. This, in essence, is the photoelectric effect.

The photons of a beam of light have a characteristic energy proportional to the frequency of the light. In the photoemission process, if an electron within some material absorbs the energy of one photon and acquires more energy than the work function of the material (the electron binding energy), it is ejected. If the photon energy is too low, the electron is unable to escape the material. Increasing the intensity of the light increases the number of photons in the beam of light and thus increases the number of electrons excited but does not increase the energy that each electron possesses. The energy of the emitted electrons does not depend on the intensity of the incoming light (the number of photons), only on the energy or frequency of the individual photons. It is strictly an interaction between the incident photon and the outermost electron.

Electrons can absorb energy from photons when irradiated, but they usually follow an all-or-nothing principle. Typically, one photon is either energetic enough to cause emission of an electron or the energy is lost as the atom returns back to the ground state. If excess photon energy is absorbed, some of the energy liberates the electron from the atom and the rest contributes to the electron's kinetic energy as a free particle.

Experimental Observations of Photoelectric Emission

For a given metal, there exists a certain minimum frequency of incident radiation below which no photoelectrons are emitted. This frequency is called the threshold frequency. Increasing the frequency of the incident beam and keeping the number of incident photons fixed (resulting in a proportionate increase in energy) increases the maximum kinetic energy of the photoelectrons emitted. The number of electrons emitted also changes because the probability that each impacting photon results in an emitted electron is a function of the photon energy. However, if just the intensity of the incident radiation is increased, there is no effect on the kinetic energies of the photoelectrons.

For a given metal and frequency of incident radiation, the rate at which photoelectrons are ejected is directly proportional to the intensity of the incident light. An increase in the intensity of the incident beam (keeping the frequency fixed) increases the magnitude of the photoelectric current, though the stopping voltage remains the same. The time lag between the incidence of radiation and the emission of a photoelectron is very small, less than 10^{-9} second, and is unaffected by intensity changes.

Applications of Photoelectric Effect

Devices based on the photoelectric effect have several desirable properties, including producing a current that is directly proportional to light intensity and a very fast response time. One basic device is the photoelectric cell, or photodiode. Originally, this was a phototube, a vacuum tube containing a cathode made of a metal with a small work function so that electrons would be easily emitted. The current released by the plate would be gathered by an anode held at a large positive voltage relative to the cathode. Phototubes have been replaced by semiconductor-based photodiodes that can detect light, measure its intensity, control other devices as a function of illumination, and turn light into electrical energy. These devices work at low voltages, comparable to their bandgaps, and they are used in industrial process control, pollution monitoring, light detection within fibre optics telecommunications networks, solar cells, imaging, and many other applications.

Photoconductive cells are made of semiconductors with bandgaps that correspond to the photon energies to be sensed. For example, photographic exposure meters and automatic switches for street lighting operate in the visible spectrum, so they are typically made of cadmium sulfide. Infrared detectors, such as sensors for night-vision applications, may be made of lead sulfide or mercury cadmium telluride.

Photovoltaic devices typically incorporate a semiconductor p-n junction. For solar cell use, they are usually made of crystalline silicon and convert about 15 percent of the incident light energy into electricity. Solar cells are often used to provide relatively small amounts of power in special environments such as space satellites and remote telephone installations. Development of cheaper materials and higher efficiencies may make solar power economically feasible for large-scale applications.

The photomultiplier tube is a highly sensitive extension of the phototube, first developed in the 1930s, which contains a series of metal plates called dynodes. Light striking the cathode releases electrons. These are attracted to the first dynode, where they release additional electrons that strike the second dynode, and so on. After up to 10 dynode stages, the photocurrent is so enormously amplified that some photomultipliers can virtually detect a single photon. These devices, or solid-state versions of comparable sensitivity, are invaluable in spectroscopy research, where it is often necessary to measure extremely weak light sources. They are also used in scintillation counters, which contain a material that produces flashes of light when struck by X rays or gamma rays, coupled to a photomultiplier that counts the flashes and measures their intensity. These counters support applications such as identifying particular isotopes for nuclear tracer analysis and detecting X rays used in computerized axial tomography (CAT) scans to portray a cross section through the body.

Photodiodes and photomultipliers also contribute to imaging technology. Light amplifiers or image intensifiers, television camera tubes, and image-storage tubes use the fact that the electron emission from each point on a cathode is determined by the number of photons arriving at that point. An optical image falling on one side of a semitransparent cathode is converted into an equivalent "electron current" image on the other side. Then electric and magnetic fields are used to focus the electrons onto a phosphor screen. Each electron striking the phosphor produces a flash of light, causing the release of many more electrons from the corresponding point on a cathode directly opposite the phosphor. The resulting intensified image can be further enhanced by the same process to produce even greater amplification and can be displayed or stored.

At higher photon energies the analysis of electrons emitted by X rays gives information about electronic transitions among energy states in atoms and molecules. It also contributes to the study of certain nuclear processes, and it plays a role in the chemical analysis of materials, since emitted electrons carry a specific energy that is characteristic of the atomic source. The Compton effect is also used to analyze the properties of materials, and in astronomy it is used to analyze gamma rays that come from cosmic sources.

References

- Jung, hye song; young joon hong, yirui li, jeonghui cho, young-jin kim, gyu-chui yi (2008). "photocatalysis using gan nanowires". Acs nano. 2 (4): 637–642. Doi:10.1021/nn700320y.cs1 maint: multiple names: authors list (link)

- Mammana, a.; et al. (2011). "a chiroptical photoswitchable dna complex". Journal of physical chemistry b. 115 (40): 11581–11587. Doi:10.1021/jp205893y. Pmid 21879715

- Berinstein, paula (2001-06-30). Alternative energy: facts, statistics, and issues. Greenwood publishing group. Isbn 1-57356-248-3

- Vachon, j.; et al. (2014). "an ultrafast surface-bound photo-active molecular motor". Photochemical and photobiological sciences. 13 (2): 241–246. Doi:10.1039/c3pp50208b. Pmid 24096390

- Steve heckeroth (february–march 2010). "the promise of thin-film solar". Mother earth news. Retrieved march 23, 2010

Permissions

All chapters in this book are published with permission under the Creative Commons Attribution Share Alike License or equivalent. Every chapter published in this book has been scrutinized by our experts. Their significance has been extensively debated. The topics covered herein carry significant information for a comprehensive understanding. They may even be implemented as practical applications or may be referred to as a beginning point for further studies.

We would like to thank the editorial team for lending their expertise to make the book truly unique. They have played a crucial role in the development of this book. Without their invaluable contributions this book wouldn't have been possible. They have made vital efforts to compile up to date information on the varied aspects of this subject to make this book a valuable addition to the collection of many professionals and students.

This book was conceptualized with the vision of imparting up-to-date and integrated information in this field. To ensure the same, a matchless editorial board was set up. Every individual on the board went through rigorous rounds of assessment to prove their worth. After which they invested a large part of their time researching and compiling the most relevant data for our readers.

The editorial board has been involved in producing this book since its inception. They have spent rigorous hours researching and exploring the diverse topics which have resulted in the successful publishing of this book. They have passed on their knowledge of decades through this book. To expedite this challenging task, the publisher supported the team at every step. A small team of assistant editors was also appointed to further simplify the editing procedure and attain best results for the readers.

Apart from the editorial board, the designing team has also invested a significant amount of their time in understanding the subject and creating the most relevant covers. They scrutinized every image to scout for the most suitable representation of the subject and create an appropriate cover for the book.

The publishing team has been an ardent support to the editorial, designing and production team. Their endless efforts to recruit the best for this project, has resulted in the accomplishment of this book. They are a veteran in the field of academics and their pool of knowledge is as vast as their experience in printing. Their expertise and guidance has proved useful at every step. Their uncompromising quality standards have made this book an exceptional effort. Their encouragement from time to time has been an inspiration for everyone.

The publisher and the editorial board hope that this book will prove to be a valuable piece of knowledge for students, practitioners and scholars across the globe.

Index

www.ingramcontent.com/pod-product-compliance
Lightning Source LLC
Chambersburg PA
CBHW082015190326
41458CB00010B/3196